# NEURONAL CELL APOPTOSIS RESEARCH

# NEURONAL CELL APOPTOSIS RESEARCH

RELIF J. SHANDLING
EDITOR

Nova Science Publishers, Inc.
*New York*

**LIBRARY OF CONGRESS CATALOGING-IN-PUBLICATION DATA**

Neuronal cell apoptosis research / Relif J. Shandling (editor).
p. ; cm.
Includes bibliographical references and index.
ISBN 13 978-1-60021-456-1
ISBN 10 1-60021-456-8
1. Neurons. 2. Apoptosis. I. Shandling, Relif J.
[DNLM: 1. Apoptosis--physiology. 2. Neurons--physiology. QU 375 N494 2006]
QP356.25.N475 2006
571.9'36--dc22 2006030117

*Published by Nova Science Publishers, Inc. ✦ New York*

# Contents

# PREFACE

Apoptosis is the regulated form of cell death. It is a complex process defined by a set of characteristic morphological and biochemical features that involves the active participation of affected cells in a self-destruction cascade. This programmed cell death plays a critical role in physiological functions such as cell deletion during embryonic development, balancing cell number in continuously renewing tissues and immune system development. Additionally, a dysregulation of apoptosis is underlying in numerous pathological situations such as Parkinson, Alzheimer′s disease and cancer. A number of studies have pointed out an association between consumption of fruits and vegetables, and certain beverages such as tea and wine, which are rich in polyphenols, with reduced risk of chronic diseases, including cancer. Apoptosis is also the regulatory mechanism involved in the removal of unnecessary cells during development and in tissue homeostasis in a wide range of organisms from insects to mammals. The aim of this book is to provide new studies in the field of neuronal apoptosis research.

Apoptosis — the programmed cell death — is a complicated process. The activity of many genes influences a cell's likelihood of activating its self-destruction program. Once "the decision" is taken, the apoptotic program, in the induction phase, requires the coordinated activation and execution of multiple events. They constitute a dedicated apoptotic signalling pathway converging into the common events of the execution phase. Oxidative stress has often been evoked as a mediator of this intracellular apoptotic signalling. Within a certain concentration range, reactive oxygen species (ROS) can induce stress response of the cells by altering expression of respiratory genes to uphold the energy metabolism to rescue the cell. However, ROS may both cause a wide spectrum of oxidative damage to various cellular components which result in cell death and elicit apoptosis by induction of release from mitochondria of apoptogenic factors including cytochrome *c*. A major role in apoptosis is also played by the antioxidant and the proteolytic cell systems.

Mitochondria play a pivotal role in apoptosis being recently emerged as a convergence centre for life and death signals.

In chapter 1, the authors discuss how ROS act to regulate apoptosis in a coordinated manner and how the released cytochrome *c* participates to apoptosis, being able to act as an antioxidant itself, as well as responsible of caspase activation, which play a prominent role in the destructive phase of apoptosis.

The role of the adenine nucleotide translocator will be also reported as a responsible for ATP export outside mitochondria and as a component of the mitochondrial permeability transition pore.

Primary rat cerebellar granule cell culture will be considered as a model cell system and will be investigated *en route* to apoptosis.

The generation of insoluble peptide/protein fibrils is well-confirmed in amyloidosis, complex disorders characterized by the polymerization and aggregation of normally innocuous and soluble proteins or peptides followed by extracellular insoluble fibrils with resistance to peptidases. In chapter 2, the authors demonstrated that pathologically-related amyloidogenic proteins/peptides, including prion protein fragment (PrP), ß-amyloid(1–42), and amylin, generated the insoluble fibrils, and it induced delayed cell death in neuronal PC-12 cells. In addition to these pathogenic peptides/proteins, some therapeutic peptides, including glucagon and salmon calcitonin, also had an ability to form gel-like fibrillar aggregates as observed by transmission electron microscopy. On the basis of circular dichroic analysis, the peptides-derived fibrils were found to share a distinct conformational feature in the richness of the ß-sheet structure, despite having no sequential homology among these peptides. All tested peptide fibrils, including the pathologically-related and therapeutic amyloidogenic proteins/peptides, caused neurotoxic effects with the significant activation of caspase-3, an apoptotic enzyme, in PC12.

Pituitary adenylate cyclase-activating polypeptide (PACAP) and vasoactive intestinal peptide (VIP), peptide transmitters in the central and peripheral nervous systems, potently stimulate neuritogenesis and survival of neuronal cells in the central nervous system. Concomitant application of these neuropeptides with peptides aggregates, such as aged PrP(106–126) and glucagon, inhibited the delayed death of neuron-like PC12 cells. In particular, PACAP27 inhibited the neurotoxicity of peptides fibrils at low concentrations ($>10^{-15}$ M), characterized by the deactivation of fibrils-stimulated caspase-3. The neuroprotective effect of PACAP27 was antagonized by the selective PKA inhibitor, H89, or MAP kinase inhibitor, U0126, suggesting that PACAP27 attenuates fibrils-induced delayed neurotoxicity in PC12 cells by activating both PKA and MAP kinases mediated by PAC1 receptor.

One of the challenges in the next few years will be to clarify the apoptotic routes in neurons in order to design potential drugs for the treatment of Alzheimer's disease, Parkinson's disease and other neurological disorders. Although an enormous effort has been made in the past few years and it has been evidenced that the mitochondria comprise a key component of the neuronal apoptotic route, it seems that in addition to the mitochondria other intracellular components are implicated in this process. With reference to this, it has been proposed that DNA damage and reentry into the cell cycle could constitute a common pathway in the apoptotic process in neurological diseases. This hypothesis is supported by the existing data in regard to neurodegenerative disorders in the brain of patients who show an increase in cell cycle protein expression compared with control brains. Indeed, studies performed in neuronal cell preparations suggest that reentry into the cell cycle and, concretely, increased expression of the transcription factor E2F1 may be the prominent pathway involved in the regulation of apoptosis. In chapter 3, the authors also summarize the potential routes involved in E2F1 induced apoptosis, p53 dependent and p53 independent. In this last hypothesis, multiple pathways have been evidenced, including the activation of caspases. Finally, given the increasing experimental data on the neuroprotective and antiapopotic effects of CDKs inhibitory drugs, among them flavopiridol, it is logical to predict the potential application of these drugs for the treatment of neurological disorders.

Programmed cell death or apoptosis is a relevant process in the physiology and pathology of the nervous system. Apoptosis is an organized form of cell death which is triggered by different factors including DNA damage. A growing body of evidence suggests that DNA damage and genomic instability are involved in neuronal abnormalities and may play a central role in neurodegeneration. DNA strand breaks and DNA lesions have been reported in Parkinson's and Alzheimer's diseases and as an early event after reperfusion of ischemic brain. DNA damage has been found to activate a cell death program in terminally differentiated postmitotic neurons. Since the genome is continuously damaged by a variety of endogenous and exogenous agents and the majority of DNA damage is produced by oxyradicals, generated by normal aerobic metabolism, neurons are particularly susceptible to DNA damage due to the high rate of oxidative metabolism. To maintain genomic integrity, cells are equipped with special defense mechanism, DNA damage response, to remove DNA damage by DNA repair pathways or eliminate damaged cells via apoptosis. Generally, differentiated cells, like neurons, are deficient in DNA repair and more vulnerable to DNA damage-initiated apoptosis. For example, neurons are more vulnerable than astrocytes to DNA-damaging conditions such as ionizing radiation. Breast cancer patients receiving hemotherapy commonly experience long-lasting cognitive impairment. It is known that radiotherapy may cause CNS toxicity. DNA damaging agents including γ-irradiation induce neuronal apoptosis *in vitro*, suggesting the direct adverse effect of these DNA damaging agents on neurons. The importance of DNA repair for neuronal survival is illustrated by disorders observed in patients with hereditary DNA repair abnormalities. These disorders combine the predisposition to cancer with progressive neurodegeneration. Although indirect evidence suggests that DNA damage and repair mechanisms play critical roles in neuronal survival, the pathways involved are poorly understood. Recently, the authors in chapter 4 have found that cell cycle activation is essential for DNA damage-induced neuronal apoptosis which suggests that the cell cycle machinery is a critical element of the DNA damage response not only in cycling but also in quiescent cells. Here, they discuss the DNA damage response in postmitotic neurons and possible mechanisms by which neurons are forced to apoptosis versus DNA repair thereby controlling cell fate. Elucidation of these mechanisms promises to provide multiple points of therapeutic intervention in neurodegenerative diseases.

Apoptosis is a key physiological response that occurs during development of the nervous system, resulting in the death of nearly half of the embryonic neuronal population, while aberrant apoptotic mechanisms are thought to contribute significantly to many neurological disorders including Alzheimer's disease. Although many experiments in the past have demonstrated the requirement of *de novo* gene expression during neuronal apoptosis, the complete spectrum of genes involved in distinct temporal domains is mostly unknown. As discussed in chapter 5, to begin a comprehensive survey of the gene-based molecular mechanisms that underlie neuronal apoptosis, the authors have used the unprecedented experimental opportunities that genome sequences and the development of DNA microarray technology now provide to perform genome-wide expression analysis in different paradigms of neuronal apoptosis. This new approach offers a genomic view of the changes that accompany neuronal apoptosis and yields new insights into the molecular basis underlying it.

With regard to the specificity of the developing brain, a better understanding of cellular mechanisms involved in perinatal hypoxic-ischemic injury would help to prevent neurological impairments. In chapter 6, the authors examined temporal features of brain injury in three different developmental models of oxygen deprivation capable of inducing apoptotic cell death. Nuclear staining by DAPI (4,6-diamidino-2-phenylindole) was used to identify healthy, apoptotic and necrotic nuclei as well as for cell counting in cultures and brain sections. DNA fragmentation was monitored by in situ terminal dUTP nick end labeling (TUNEL) and electrophoresis on agarose gels. Also, the expression profile of apoptosis-related proteins Bax and Bcl-2 was studied by immunohistochemistry. In all cases, oxygen deprivation induced significant delayed cell death with morphological features of apoptosis and a progressive increase in the Bax/Bcl-2 protein ratio, except in the penumbra of the ischemic infarct where Bcl-2 remained predominant. As in the control newborn brain that still exhibited physiological death, hypoxia-associated DNA breakdown led to small fragments of ~200 bp in the cortex of hypoxic rat pups. Ladder pattern and TUNEL-positive cells exhibiting apoptotic bodies were only present in the penumbra of 7 day-old ischemic rats. These data indicate that hallmarks of hypoxia-induced apoptosis may vary according to brain maturity, possibly through specific nuclease activities. While retaining a part of the developmental death program, the newborn brain seems to be prone to an apoptotic-like response that resembles physiological programmed death.

Apoptosis, known as programmed cell death, is a form of cell death that serves to eliminate dying cells in proliferating or differentiating cell populations, thus plays a crucial role in normal growth and development of many tissues including neurons and tissue homeostasis. Brain development begins in intrauterine period and continues until the end of the adolescence. Developing brain is much more sensitive to environmental exposures such as maternal deprivation, stress, ethanol, head trauma, exercise. Apoptosis plays an important role in developing brain and also environmental exposures can cause apoptosis or increase apoptosis.

Many neuroprotective agents such as melatonin, erythropoietin, MK-801 may prevent environmental exposures-induced apoptosis in developing brains. As discussed in chapter 7, approximately on postnatal day 4, the infant rat enters stress-hyporesponsive period (SHRP). A single episode of maternal deprivation for 24h at post-SHRP induce neuronal loss in hippocampus regions of the brain in the infant rats, while it is not influence hippocampal neurons in SHRP. Environmental exposures such as stress, not affect hippocampal neurons in this period. Infant brain might be protected in SHRP from stress-induced apoptosis of brain. Also, head trauma activated physiological apoptotic mechanisms in the developing brains. The immature brain is exceedingly vulnerable to apoptotic neurodegeneration triggered by head trauma and ethanol intake. The effects of exercise opposite of these exposures, increase cell density without altering of apoptosis in the hippocampus and dentate gyrus of adolescent rats. Hippocampal neurons normally play a crucial role in the processing of spatial memory and learning. Memory function, in particular, is at risk following environmental exposures-induced apoptosis due to the selective vulnerability of the hippocampus. To keep a constant number of cells in tissues, there has to be a balance between new cell formation (or proliferation) and cell death (apoptosis). Hippocampal neuron loss is associated with memory deficits following apoptosis. Likewise, increase of hippocampal neurons is closely related to increase in learning and memory.

Apoptosis is an important process in the development of the nervous system, and also apoptosis may be alterable extrinsic factors.

Trying to protect ecosystems and humans as well as trying to answer several questions about the effect, transport and fate of natural and synthetic toxicants, ecotoxicology uses tools and methods from different disciplinary areas and resorts to biological markers indicative of exposure to or effects of toxicants. It is known that cellular death can be promoted by several toxicants and pathological conditions. Generally, when the concentration of a toxicant is high enough to cause irreparable cellular injuries or large disorders in the cellular milieu the cell dies by necrosis, through a process without energy consumption, but on the other hand, it is much more common affected cells to die by apoptosis, which consumes energy.

Apoptosis is a highly conserved and regulated mechanism, essential to organ development, homeostasis and function. This mechanism involved in cell turnover accounts for normal and pathological modifications of tissues, it may occur spontaneously in malignant tumors, and its induction or suppression in the beginning of disease can have huge impact on tissue and organ performance. Additionally, apoptosis can be a consequence of the action of several toxicants, inorganic and organic, present in the air, food, and water to which living organisms are exposed. Trace metals, such as cadmium, mercury, copper, and zinc, individually or in mixtures, are known to induce apoptosis when biological systems are exposed within certain concentration limits. The same phenomenon can occur when organisms are exposed to organic compounds, such as polcycyclic aromatic hydrocarbons (PAHs), polychlorinated biphenyls (PCBs), dioxins, and other persistent pollutants. Apoptosis, which is expressed by specific molecular and cellular morphological features, is therefore an essential issue of research in ecotoxicology since it is strongly correlated to the action of many toxicants and can be important as a biomarker of such action. Thus, apoptosis and its use as a biomarker of inorganic and organic toxicants are here addressed in an ecotoxicological approach. The benefit of apoptosis detection in biomonitoring programs is also emphasized in chapter 8.

In: Neuronal Cell Apoptosis Research
Editor: Relif J. Shandling, pp. 1-46
ISBN 1-60021-456-8


*Chapter 1*

# THE MITOCHONDRIA IN CEREBELLAR GRANULE CELLS *EN ROUTE* TO APOPTOSIS

***A. Atlante[1], A. Bobba[1], E. Marra[1] and S. Passarella[2]***
[1]Istituto di Biomembrane e Bioenergetica, CNR, Via G. Amendola, 165/A - 70126 Bari, Italy;
[2]Dipartimento di Scienze per la Salute, Università del Molise, Via A. De Sanctis - 86100 Campobasso - Italy.

## ABSTRACT

Apoptosis — the programmed cell death — is a complicated process. The activity of many genes influences a cell's likelihood of activating its self-destruction program. Once "the decision" is taken, the apoptotic program, in the induction phase, requires the coordinated activation and execution of multiple events. They constitute a dedicated apoptotic signalling pathway converging into the common events of the execution phase. Oxidative stress has often been evoked as a mediator of this intracellular apoptotic signalling. Within a certain concentration range, reactive oxygen species (ROS) can induce stress response of the cells by altering expression of respiratory genes to uphold the energy metabolism to rescue the cell. However, ROS may both cause a wide spectrum of oxidative damage to various cellular components which result in cell death and elicit apoptosis by induction of release from mitochondria of apoptogenic factors including cytochrome *c*. A major role in apoptosis is also played by the antioxidant and the proteolytic cell systems.

Mitochondria play a pivotal role in apoptosis being recently emerged as a convergence centre for life and death signals.

Here we discuss how ROS act to regulate apoptosis in a coordinated manner and how the released cytochrome *c* participates to apoptosis, being able to act as an antioxidant itself, as well as responsible of caspase activation, which play a prominent role in the destructive phase of apoptosis.

The role of the adenine nucleotide translocator will be also reported as a responsible for ATP export outside mitochondria and as a component of the mitochondrial permeability transition pore.
Primary rat cerebellar granule cell culture will be considered as a model cell system and will be investigated *en route* to apoptosis.

# 1. APOPTOSIS: GENERAL ASPECTS

People consider that cell death can occur essentially via either necrosis or apoptosis. In the terms of Kerr *et al.* [1972], apoptosis describes "the orchestrated collapse of a cell characterised by membrane blebbing, cell shrinkage, condensation of chromatin, and fragmentation of DNA followed by rapid engulfment of the corpse by neighbouring cells" (Figure 1). It is distinguished from necrosis owing to the absence of any inflammatory response. Moreover in distinction from apoptosis, necrosis exhibits L-lactate dehydrogenase release and deficit in ATP production. At present it has been recognised that apoptosis takes place in a variety of biological processes sometimes opposite including embryogenesis (see the interdigital webs present during early mammalian embryonic life) and ageing, normal tissue homoeostasis and diseases. This is why apoptosis has become one of the hottest fields of biomedical research.

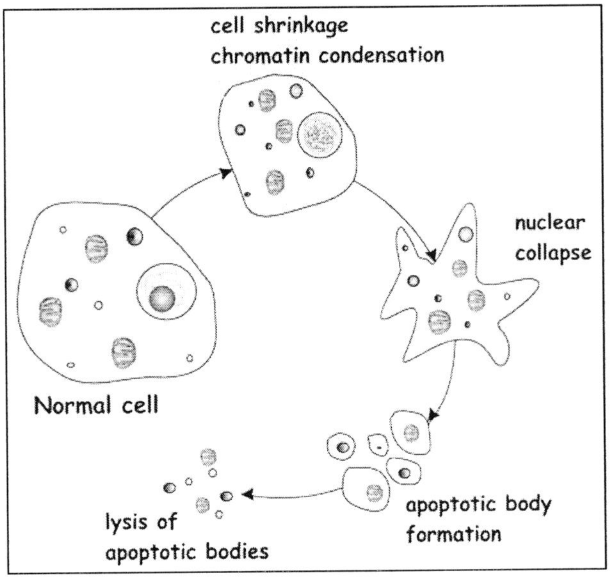


Figure 1. How apoptosis occurs.

We will assume that cells die via apoptosis when a number of specific hallmarks are fulfilled. They are chromatin condensation, DNA fragmentation, phosphatidylserine externalization, blebbing of the plasma membrane, cell shrinkage and importantly, the expression of a variety of genes with the consequent possibility to prevent death by inhibiting their expression. Indeed apoptosis is a process in which cells play an active role in their own death (cell suicide). Programmed cell death will refer to other forms of cell death that require gene expression without fulfilling some, or all, of the other criteria of apoptosis [Sperandio *et al.* 2000].

We suggest to distinguish in the experimental and clinic studies at least three different steps for cells *en route* to apoptosis, being each separate from the others: they are initiation, genetic regulation, and effector mechanisms [Renehan, 2001]. Initiators of apoptosis include anticancer drugs, gamma and ultraviolet irradiation, deprivation of survival factors such as interleukin- and various other cytokines that activate "death receptors" and tumour necrosis factor receptors. Through a variety of pathways these stimuli, by eliciting effectors, in turn, generate a characteristic pattern of gene expression. For instance, the members of the Bcl-2 gene family can either promote or suppress cell death by modulating mitochondrial function. Some of these proteins (Bcl-2 and bcl-$X_L$ etc) are anti-apoptotic, while others (Bad, Bax etc) are pro-apoptotic. The sensitivity of cells to apoptosis depends on the balance of pro- and anti-apoptotic bcl-2 proteins [Tsujimoto and Shimizu, 2000]. Then, when there is an excess of pro-apoptotic proteins the cells are more sensitive to apoptosis, when there is an excess of anti-apoptotic proteins the cells will tend to be more resistant (Figure 2).

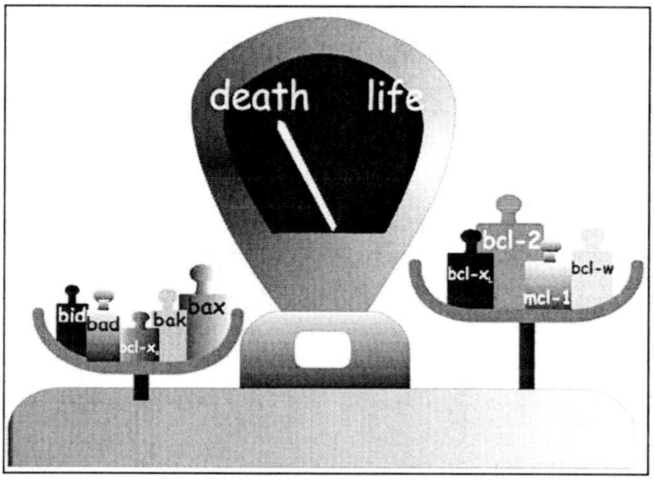


Figure 2. The balance between cell survival and cell death due to the components of the Bcl-2 gene family.

A variety of extracellular signals and intracellular mediators is also involved in maintaining this balance. When the cell is exposed to physical, biochemical or biological injury, or deprived of necessary substances, a series of stress-response genes is activated. With minimal insults, the cell may recover. With greater insults, apoptosis takes place; the cell dies and is recycled to its neighbours. Dys-regulation of the physiological control of cell homeostasis results in disease [Evan and Littlewood, 1998; Kroemer and Reed, 2000; Reed, 2000]. For example, in cancer many cells have lost the ability to undergo apoptosis in response to normal signals or after damage from therapeutic interventions [Arceci, 1996; Guchelaar *et al.*, 1997], on the other hand unrestrained apoptosis of neurons may contribute to Alzheimer's and other neurodegenerative diseases [Gorman *et al.*, 1996; Deigner *et al.*, 2000; Canu and Calissano, 2003]. The principal effectors are a family of proteases termed caspases.

Among the cell components, mitochondria play a major role in apoptosis.

## 1.1. Mitochondria in Apoptosis: the Cell's Powerhouse and what Else?

Mitochondrion is a dynamic organelle which defines in the cytoplasm a "close space" due to its compartments: the outer membrane (OMM), the intermembrane space (IMS), the inner membrane (IMM) and the matrix space. Notice that recently other compartments have been defined in the mitochondria: cristal membranes and intracristal space [Logan, 2006]. The results obtained using advanced tomographic imaging techniques demonstrate that, at least in animal tissue, tubular rather than lamellar cristae predominate and that the morphology of cristae infers that they are structurally distinct from the rest of the inner mitochondrial membrane. An additional finding was confirmation that the cristae were connected to the inner membrane (cortical inner mitochondrial membrane, parallel to the outer membrane) by membranous tubules, instead of the cristae being simple in-folds of the membrane as suggested by Palade [1952].

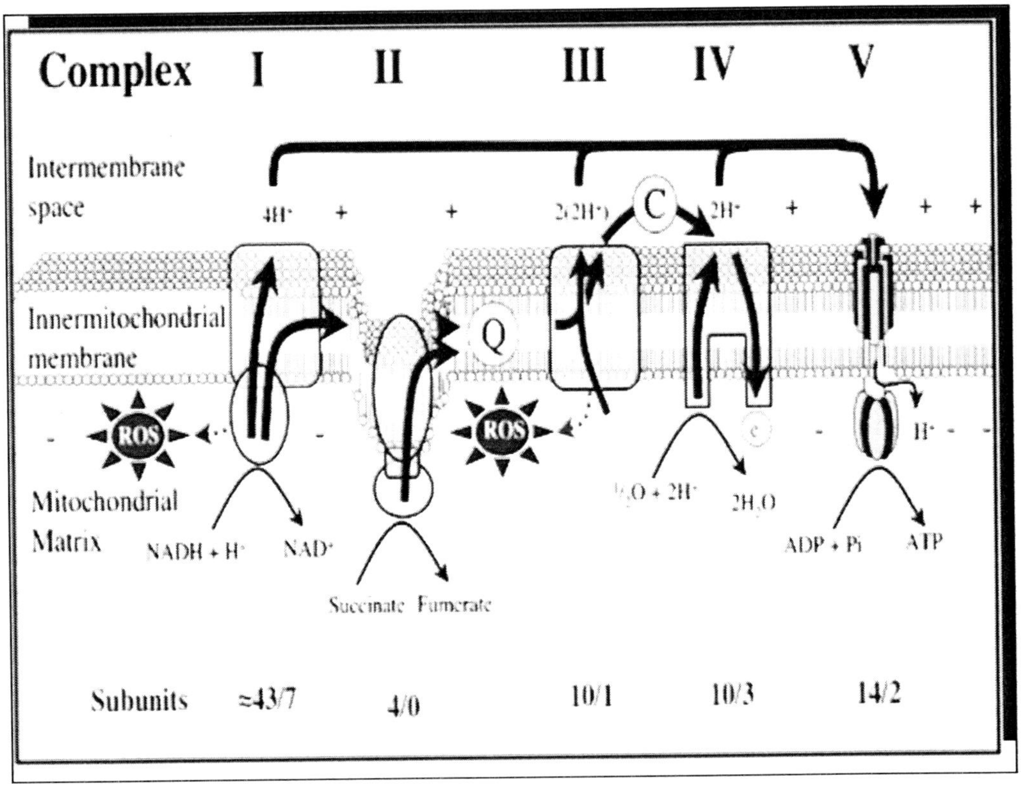


Figure 3. Oxidative phosphorylation (from: *www.myweb.uiowa.edu/bballard/OXPHOS.htm*).

Mitochondria undergo drastic changes in volume and state of membrane organization during electron flow along the respiratory chain. In the matrix, respiratory substrates are oxidized by their specific dehydrogenases, mainly through the tricarboxylate cycle and the β-oxidation pathway, thus generating NADH and $FADH_2$. The mitochondrial respiratory chain, located in the inner mitochondrial membrane, which accounts for the oxidative phosphorylation, is composed of five enzyme complexes: complex I (NADH:ubiquinone oxidoreductase), complex II (succinate: ubiquinone oxidoreductase), complex III (ubiquinol: cytochrome *c* oxidoreductase), complex IV (cytochrome *c* oxidase), complex V ($H^+$-translocating ATP synthase) and two mobile components, ubiquinone and cytochrome *c* (cyt *c*), that transport "reducing equivalents" from respiratory substrates to molecular oxygen, down the redox potential. This hydrogen/electron current forms three cascades in which the redox energy is high enough to be utilized to extrude protons from the matrix, to the intermembrane space. The electrochemical proton gradient thus formed, also designated as the protonmotive force, $\Delta\mu_H^+$, according to Mitchell's chemiosmotic theory, is the driving force for the back flow of protons through the $F_o$-$F_1$-ATP synthase complex, $\Delta\mu_H^+$ being the coupling vehicle between substrate oxidation and ADP phosphorylation to ATP catalysed by complex V (ATP synthase) (Figure 3). This gradient is composed of the electric component ($\Delta\Psi$) and the proton concentration gradient ($\Delta$pH). As a result, mitochondria are unique cellular organelles that can build up a transmembrane electric potential of up to 180 mV, negative inside, and whose internal milieu maintains a pH value of about 8 [Nicholls and Ferguson, 2002].

Experimentally, if mitochondria are incubated in an oxygraph apparatus (oxygen electrode) in an isotonic medium containing substrate and phosphate, then addition of ADP causes a sudden burst of oxygen uptake as the ADP is converted into ATP.

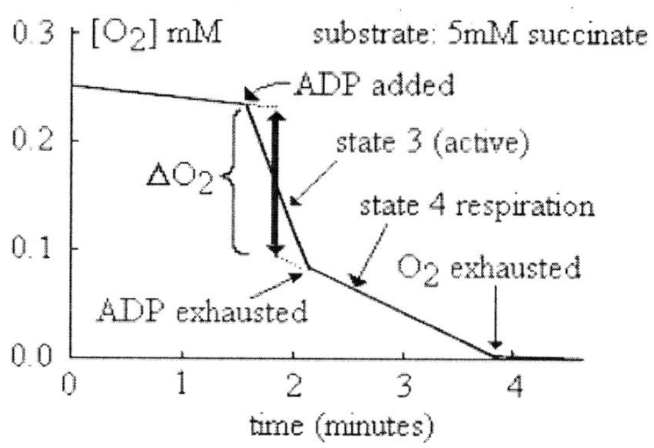


The actively respiring state is sometimes referred to as *"state 3"* respiration, while the slower rate after all the ADP has been phosphorylated to form ATP is referred to as *"state 4"*.

*State 4* respiration is usually faster than the original rate before the first addition of ADP because some ATP is broken down by ATPase activities contaminating the preparation, and the resulting ADP is then re-phosphorylated by the intact mitochondria.

The ratio [*state 3* rate]: [*state 4* rate] is called the respiratory control index (RCI) and indicates the tightness of the coupling between respiration and phosphorylation. With isolated mitochondria the coupling is not perfect, probably as a result of mechanical damage during the isolation procedure. Typical RCI values range from 3 to 10, varying with the substrate and the quality of the preparation.

Although the end product of the respiratory chain is water that is generated in a four-electron reduction of molecular oxygen by cytochrome oxidase (complex IV), a minor

unknown mechanisms (for procaspases-2 and -8) [Samali *et al.*, 1998; Susin *et al.*, 1999b; Qin *et al.*, 2001].

In summary, multiple factors function in conjunction and in parallel to trigger cell death. The release of cyt *c* activates caspases, and the release of EndoG and AIF induces DNA fragmentation and chromatin condensation. The pathway is able to feed-forward and amplify the apoptotic signal. Active caspases can leave the Bcl-2 family of proteins to cause more mitochondrial damage [Cheng *et al.,* 1997]; and active DNases will generate DNA breaks that are signals for mitochondrial damage.

Notice that the above reported conclusion remain valid in the experimental system used. Whether their function in general for all the apoptosis cases remains to be established.

Nonetheless, it remains to be fully established how the release of these proteins from the mitochondria occurs.

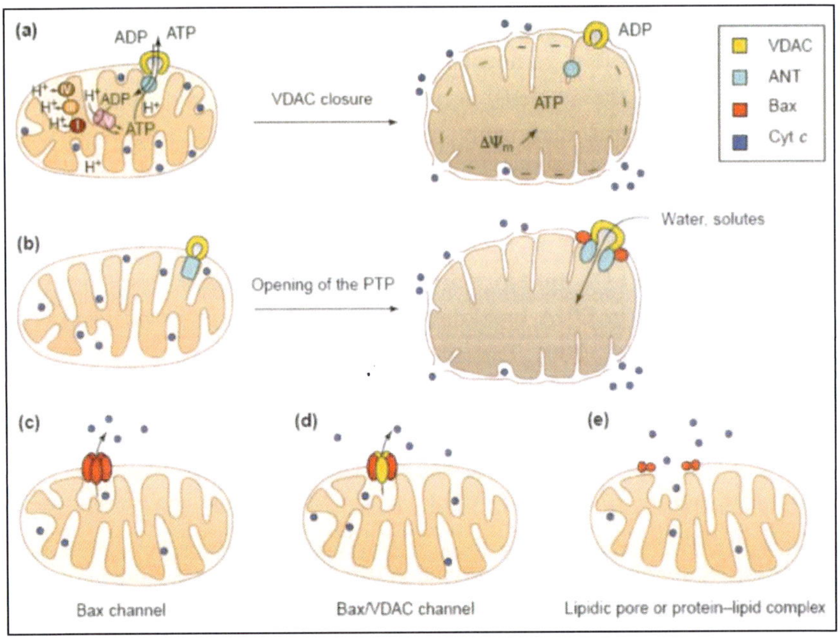


Figure 4. Five models for the release of cytochrome *c* from mitochondria during apoptosis. Reprinted from Trends in Cell Biology, Vol. 10, Desagher S and Martinou JC, *Mitochondria as the central control point of apoptosis*, 369-377, Copyright (2000), with permission from Elsevier.

For instance, five models for the release of cyt *c,* as well as the other mitochondrial proteins, from mitochondria during apoptosis have been proposed by Desagher and Martinou [2000] (see Figure 4). Model (*a*) involves closure of the voltage-dependent anion channel (VDAC) and impairment of ATP–ADP exchange [Vander Heiden, 1999]. Under normal conditions (left), the protons that accumulate in the intermembrane space as a result of electron transport re-enter the mitochondrial matrix through F1F0ATPase (complex V of the respiratory chain). By inhibiting F1F0ATPase, impairment of ATP–ADP exchange results in the accumulation of protons in the intermembrane space and an increase in the mitochondrial transmembrane potential ($\Delta\Psi$ m), which is predicted to promote osmotic swelling (right). The outer mitochondrial membrane ruptures as a result of swelling of the mitochondrial matrix, allowing cyt *c* and other proteins to escape from the intermembrane space. Model (*b*) suggests that the permeability-transition pore (PTP) opens and that the inner membrane permeability increases, causing mitochondrial-matrix swelling [Kroemer *et al.*, 1997]. Bax might induce this pore to open by binding to the adenine-nucleotide translocator (ANT) at the contact sites between the inner and outer membranes or after translocation to the inner membrane [Marzo *et al.*, 1998]. Models (*c*–*e*): a large channel forms in the outer mitochondrial membrane, allowing cyt-*c* release, but mitochondria are not damaged. This channel could be formed by Bax only [Antonsson *et al.*, 1997] (*c*), Bax in association with VDAC [Shimizu *et al.*, 1999] (*d*) or could be a lipidic channel or protein–lipid complex (*e*) formed after Bax insertion into the outer mitochondrial membrane [Basanez *et al.*, 1999]. Nonetheless, evidence is emerging that the integration and biological outcome of these pathways might be critically dependent on the unique lipid composition of mitochondrial membranes [Malisan and Testi, 2003].

The caution above arises to a general undefined generalization of the apoptosis mechanism, applies also for this process.

During apoptosis, mitochondria interact with other cell reagents or supramolecular complexes, including ROS, apoptosome and the proteolytic systems: proteasome and caspases.

## 1.3. Cell Participants in Apoptosis

Reactive oxygen species are amongst the most potent and omnipresent threats faced by any living organism. Intracellular accumulation of ROS such as superoxide anion, hydrogen peroxide, singlet oxygen, hydroxyl radical, and peroxy radical, can arise from toxic insults and normal metabolic processes. These species in the absence or in the minor efficiency of the cell's natural antioxidant systems, can damage all of the major classes of biological macromolecules, including nucleic acids, proteins, carbohydrates, and lipids.

Oxidative stress has been defined as a disturbance in the prooxidant-antioxidant balance, resulting in potential cell damage [Harman, 1992]. An excess of ROS causes an oxidative stress, with an enhanced activity of the antioxidant defence system and cell damage. A role for oxidative stress in apoptosis has been shaped by several independent observations. For many years, direct treatment of cells with oxidants like hydrogen peroxide or redox-active quinones was thought to exclusively cause necrosis, but more recent studies have shown that lower doses of these agents can trigger apoptosis [see below]. Numerous, recent studies have shown that the mode of cell death depends on the severity of the insult. Oxidants and antioxidants can not only determine cell fate, but also how cell dies. Interestingly, as a function of the level of ROS, cerebellar granule cells can die via apoptosis, via necrosis, or as a result of "energy catastrophe" [Atlante *et al.,* 2003a].

However, ROS are not only dangerous molecules for the cell, but they also display a physiological role, as mediators in signal transduction pathways, activating proteins such as tyrosine kinase, mitogen-activated protein kinase system or small Ras proteins [Fleury et al., 2002] (see below).

Although ROS can injure biological molecules, such as DNA, proteins and lipids, causing cell and tissue damage leading to aging and disease, the concerted action of the various antioxidants, acting as scavengers of ROS, keeps concentration of free radicals in cells relatively low. The role of antioxidant enzymes is interpreted in terms of fine tuning of the concentration of reactive oxygen species which are required in the redox regulation of the cell cycle and of programmed cell death.

Cells contain three endogenous antioxidant enzymes: superoxide dismutase, catalase and glutathione peroxidase. They can act as pro-apoptotic inducers in some cell types, such as smooth muscle cells, and as anti-apoptotic agents in other models [for refs. see Kahl *et al.*, 2004]. Interestingly, MnSOD was suggested to be inhibitor of apoptosis, infact increased MnSOD activity has been shown to prevent cell death via the receptor-mediated apoptotic pathway as well as cell death via the mitochondrial pathway [Fujimura *et al.*, 1999; 2000]. The role of both catalase and glutathione peroxidase is rather poor. Evidence for both an antiapoptotic and a proapoptotic role of catalase can be found; overexpression of glutathione peroxidase [Nakagawa, 2004a] in mammalian cells suppressed the apoptosis.

The proteasome is the major proteolytic system in the cytosol of eukaryotic cells endowed with multiple activities, referred to as chymotrypsin-like, trypsin-like, and postacidic or caspase-like activities [Peters, 1994; De Martino and Slaughter, 1999]. In the ubiquitin-proteasome pathway, proteins are first modified by the covalent conjugation to multiple ubiquitin molecules that subsequently tag the conjugates for rapid hydrolysis by the proteasome. This complex hydrolyzes exhaustively target proteins releasing small peptides, most of which are further degraded by cellular exopeptidases, with concomitant recycling of ubiquitin molecules by deubiquitinating enzymes [Wilkinson, 1997]. The inability of removing ubiquitin conjugates may result from an altered proteolytic degradation or from a modification of damaged proteins that makes them inaccessible to proteolytic machinery. Previous studies have attributed to the proteasome a critical role in apoptosis triggered by different stimuli in terminally differentiated cells [Grimm *et al.*, 1996; Sadoul *et al.*, 1996]. It has been recently reported that the proteasome system acts upstream of caspase activation in programmed cell death [Canu *et al.*, 2000; Canu and Calissano, 2003].

The apoptosome is a wheel-like structure that contains 7 molecules each of Apaf-1, cyt *c* and ATP. Cyt *c* released from the mitochondria binds to its adaptor molecule, namely apoptotic protease activating factor-1 (Apaf-1). Such an interaction results in a conformational change in Apaf-1 which, when stabilised by the binding of ATP, allows molecules of Apaf-1 to associate with each other. This wheel-like structure permits the recruitment of 7 molecules of procaspase-9 to the complex [Cain *et al.*, 1999]. The mechanism of caspase activation is still uncertain with two possibilities already proposed. In the former case, the Apaf-1, cyt *c* and procaspase-9 complex can act as a stage to activate cytosolic procaspase-9 as it is recruited to the apoptosome. In the latter, two apoptosomes have been proposed to interact with each other and to activate the caspase-9 located on the other apoptosome. Upon activation of caspase-9 within the apoptosome, the death signal is propagated through the stepwise activation of additional downstream caspases [Slee *et al.*, 1999; 2000; 2001].

Thus, upon assembly of apoptosome, rapid amplification of the death signal then occurs through activation of a panoply of other caspases that proceed to dismantle the cell.

The caspases are a group of cysteine proteases that exist within the cell as inactive pro-forms or zymogens. These zymogens can be cleaved to form active enzymes following the induction of apoptosis. Caspases are one of the main effectors of apoptosis. The finding that the class of death-related genes expressing caspases [Nicholson, 1999] is involved in cell death in genetically distant organism has directed attention on the central role of this protease family as the main effector of apoptotic cell demise. Consequently there is a tendency to identify apoptosis conceptually with its main execution system. However several forms of demise are caspase independent [Sarin *et al.*, 1997] or even accelerated by caspase inhibitors [Vercammen *et al.*, 1998a; 1998b]. In some cases, caspase inhibition does not alter the extent of death, but rather the shape of demise [Hirsch *et al.*, 1997; Leist *et al.*, 1997]. Activation of caspases has been demonstrated to occur in four separate ways; self-cleavage or cleavage by other caspases [Stennicke *et al.*, 1998; Slee *et al.*, 1999], cleavage by the granule protein granzyme B [Harvey *et al.*, 1996; Darmon *et al.*, 1996; Martin *et al.*, 1996; Van de Craen *et al.*, 1999], recruitment to ligated death receptor complexes [Medema *et al.*, 1997; Scaffidi *et al.*, 1999] or via cyt *c*-mediated activation of the apoptotic protease activating factor 1 (Apaf-1) apoptosome [Zou *et al.*, 1999; Saleh *et al.*, 1999; Wolf *et al.*, 2001]. In the latter case, cyt *c* is released into the cytoplasm (see above) where it interacts with dATP, Apaf-1 and caspase-9 (the apoptosome). This results in activation of caspase-9 that in turn activates downstream or executioner caspases. These caspases can then activate other caspases in a cascade. This cascade eventually leads to the activation of the effector caspases, such as caspase 3 and caspase 6. These caspases orchestrate the apoptotic death of a cell by cleaving a variety of key cellular proteins that leads to the typical morphological changes observed in cells undergoing apoptosis [Waterhouse *et al.*, 2002]. The degree of caspase activation may also be of importance. Caspase activation varies quantitatively from system to system, and it is unclear whether a threshold has to be reached to turn caspase activation into a lethal reaction. For example, a moderate degree of caspase activation does not always cause cell demise [Jaattela *et al.*, 1998]. Finally, an apoptotic-like death can also occur in a caspase-independent manner [see Borner and Monney, 1999]. Part of the problem may be due to the fact that caspase inhibitors can prevent the appearance of certain, but not all, features of apoptosis in certain model systems [McCarthy *et al.*, 1997]. In this context, caution should be exercised, because of possible non-specific effects of broad-range caspase inhibitors. Although generally considered specific for the caspase family, these inhibitory peptides may in fact block only some, but not all, caspases. It is also clear that caspase inhibitors can also inhibit unrelated proteases, whose activity may be necessary for survival. While the occurrence of a caspase-independent apoptosis is strongly debated, it is instead widely accepted that, in some cases, caspase inhibitors only delay cell demise. Cells die eventually with morphologically different characteristics [Hirsch *et al.*, 1997; Leist *et al.*, 1997]. Evidence that cells triggered to undergo apoptosis die by necrosis when energy levels are rapidly compromised has been recently provided [Leist *et al.*, 1997]. *In vivo*, under pathological conditions, apoptosis and necrosis may often coexist [Leist *et al.*, 1995]. Moreover, intracellular energy levels are rapidly dissipated in necrosis, but not in apoptosis [Ankarcrona *et al.*, 1994; Atlante *et al.*, 2005]. By manipulating the intracellular adenosine triphosphate (ATP) level, it has been possible to determine that when intracellular ATP is low, typical apoptotic stimuli cause necrosis [Leist *et al.*, 1997]. Lack of ATP would prevent the resulting downstream degradative processes including caspase-3 activation [Nicotera and Melino, 2004].

Due to the huge numbers of investigations in the apoptosis in which the experimental systems are different one from another and the apoptosis features are shown at different times after the apoptosis induction, a general scenario including events to be considered common to any case is still lacking. Thus, having the role of mitochondrion as the main subject to be dealt with, we will describe how apoptosis occurs as a function of time and as determined by the specific action of selected cell substrate and/or systems. We will discuss *i)* how ROS participate in apoptosis and how their level can vary in the apoptosis time by virtue of the systems devoted to their production and modulation, *ii*) how cyt c release occurs and what is the function of the released cyt c, *iii*) the role of ATP production in apoptosis as dependent on both oxidative and substrate level phosphorylation in glycolysis, *iv*) the role of caspase in causing apoptosis, in causing release of mitochondrial proteins or in altering the mitochondrial function, particularly the ATP export via the ANT. *v*) Finally the permeability transition pore will be discussed to find out whether it is dispensable or not for the apoptosis to occur.

Primary rat cerebellar granule cell (CGC) culture will be considered as a model cell system. Moreover the use of isolated coupled mitochondria as well as of cell homogenates containing intact coupled mitochondria will be considered as a tool to investigate in situ the apoptosis bioenergetics, and in particular, to make clear the role of mitochondria in the cell death.

## 2. A Versatile Cell Model System to Study Apoptosis: Cerebellar Granule Cells in Culture. Apoptosis Induction by Low Potassium

Although to die is generally an undesired affair, it has become clear that also neurons are actively programmed to commit molecular hari-kari.

Why should a neuron purposely slit its own throat and commit cellular suicide? Apoptosis is a natural mechanism to eliminate the unwanted neurons without making as big a molecular mess as doing it via necrosis. In the central nervous system (CNS), developmental apoptosis plays a major role in eliminating excess neurons and neurons useless for the synaptic connections [Oppenheim, 1991; Sadoul, 1998; Haeberlein, 2004]. In fact, in some neuronal lineages, a majority of the neurons produced die during embryogenesis. Apoptosis also plays a role in the maintenance of tissue homeostasis in the adult CNS [Polster and Fiskum, 2004]. A certain degree of neuronal cell production from neural stem cells and progenitors in localized areas of the brain is now recognized to occur also in adult life, and the continued production must be balanced by the elimination of a comparable number of existing cells to maintain homeostasis [Temple, 2001].

In this section we will describe how apoptosis essentially occurs by using cerebellar granule cells as a model system. In order to favour the understanding of the general issue, mention will be also made of other systems.

Due to their postnatal generation and the feasibility of well characterized primary *in vitro* cultures, CGCs provide a good model for the study of cellular and molecular correlates of mechanisms of survival/apoptosis and neurodegeneration/neuroprotection.

Primary cultures of CGC have been established almost 30 years ago [Gallo *et al.*, 1982; Thangnipon *et al.*, 1983; Kingsbury *et al.*, 1985] and became the most popular *in vitro* model to study most of the aspects of developmental, functional and pathological neurobiology in a rather homogenous population of neurons.

Apoptotic death of these neurons during *in vivo* development could be mimicked by manipulating various parameters in the culture condition. Calissano *et al.* [1993] identified the first *in vitro* model for apoptosis in the central nervous system, demonstrating that CGCs, deprived of depolarizing levels of extracellular potassium, undergo apoptosis [D'Mello *et al.*, 1993; Alavez *et al.*, 2003]. In fact, the survival of cells is assured if dissociated CGCs from early postnatal rats are maintained in serum-containing medium in the presence of elevated potassium levels (25mM). Depolarization due by using low potassium (5 mM) medium is assumed to mimic endogenous excitatory activity with survival promotion mediated by increases in intracellular calcium level [for refs. see Contestabile, 2002]. Dissociated CGCs develop characteristics of mature cerebellar granule cells *in vivo*, including an extensive neuritis network, expression of excitatory aminoacids receptor and production and release of L-glutamate Then, cultured from early postnatal rats, cerebellar granule cells differentiate, acquiring several morphological, biochemical, and electrophysiological characteristics of mature neurons [Levi *et al.*, 1984; Hockberger *et al.*, 1987; Cull-Candy *et al.*, 1988].

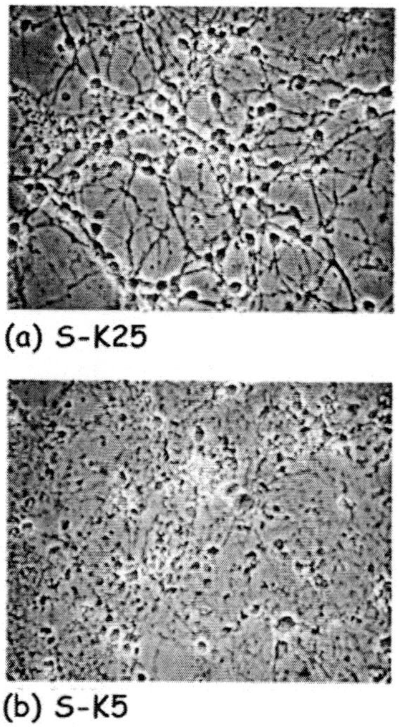


Figure 5. Phase-contrast photomicrographs of cultures of cerebellar granule cells *(a)* developed for 6-7 days in vitro in culture medium with 10% serum and 25 mM KCl (S-K25) and *(b)* subjected to potassium shift to 5 mM (S-K5) for 24 hours. Magnification 320X. Images were digitized with a Nikon TS-100 phase-contrast microscope equipped with a Nikon Digital Video Camera E995 (Coolpx 995).

To investigate the effect of low $K^+$, neurons develop for 6-7 days in culture medium with 10% serum and 25 mM KCl [Levi *et al.,* 1984; Gallo *et al*., 1987; Contestabile, 2002]: they presents round cell bodies and a complex network of neuritis (Figure 5*(a)*), which are absent in cells just plated (Figure 5*(b)*). These cells will be referred to as S-K25 cells, while the cells subjected to potassium shift to 5 mM as S-K5 cells.

The morphological changes occurring during low $K^+$-induced death in the presence of serum were identical to those seen in serum-free cultures. The death process is accompanied by cleavage of genomic DNA into internucleosome-sized fragments (Figure 6), a hallmark of apoptosis.

More importantly, DNA fragmentation was also evident in $K^+$-deprived neurons maintained in serum-containing medium. These data show that cell death is not due to serum deprivation and that lowering of $K^+$ alone is sufficient to induce apoptosis.

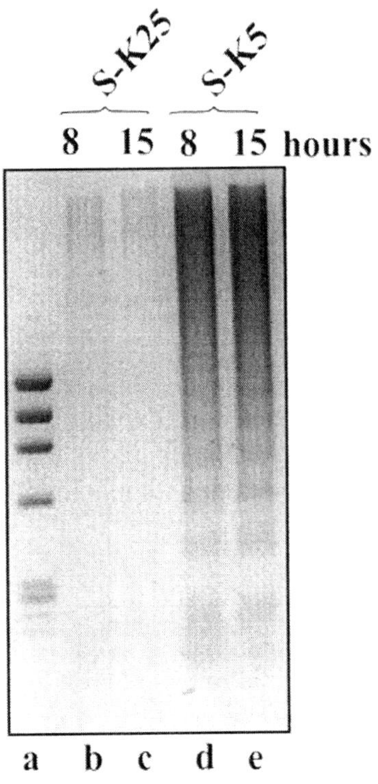


Figure 6. DNA fragmentation in cerebellar granule cells undergoing apoptosis. Lanes *b-c* and *d-e* contain DNA from control cells maintained in high potassium (S-K25) and cells switched to serum-free culture medium containing low potassium (S-K5) for 8 and 15 hours, respectively. Size marker was *HaeIII*-digested Φx174 phage DNA (lane *a*).

Exposure to actinomycin D or cycloheximide, inhibitors of mRNA and protein synthesis, at the time of $K^+$ deprivation protected against low $K^+$-induced neuronal death, this confirming that death occurs via apoptosis [D'Mello *et al.,* 1993]. It is tempting to speculate that the mechanisms involved in low $K^+$-mediated apoptosis of granule neurons are similar to those operating during neuronal death *in vivo* during development or following a blockade of neuronal activity and *in vitro* by neurotrophic factor deprivation. These mechanisms may also be activated in certain neurodegenerative diseases and during normal aging. Depolarizing concentrations of potassium are assumed to mimic *in vitro* the first synaptic afferences received *in vivo* by the cerebellar granule neurons from the mossy fibers. *In vitro* high $K^+$ is known to promote differentiation in cerebellar granule neurons, measured in terms of $K^+$-evoked glutamate release [Gallo *et al.*, 1982; Didier *et al.,* 1989], elevated glutaminase activity [Moran and Patel, 1989], establishment of mature synapses [Didier *et al.*, 1992], and regulation of the functional expression of glutamate receptors [Balàsz *et al.*, 1992; Resink *et al.*, 1994] and of glucose transporters [Maher and Simpson, 1994]. In cortical neurons, the survival-promoting effect induced by $K^+$ depolarization is attributable to the enhanced expression of brain-derived neurotrophic factor mRNA [Ghosh *et al.*, 1994]. The $K^+$ effects are mediated through influx of $Ca^{2+}$ into the cytosol, and depolarization results in a fundamental alteration in $Ca^{2+}$ homeostatic mechanisms [Tymianski *et al.*, 1994; Franklin and Johnson, 1992].

In view of the large numbers of granule neurons that can be homogeneously cultured, these results offer the prospect of an excellent model system to study the mechanisms underlying apoptosis in the central nervous system [Calissano *et al.,* 1993]. The apoptotic process consequent to the shifting of mature granule neurons to low potassium conditions, which is usually accompanied by serum deprivation, occurs in a relatively short time (neuronal death reaching 30-50% after 16-24 hours) (Figure 7), it is reproducible and, therefore, very useful for studies aimed at identifying cellular signals activated in apoptotic neurons as well as at testing ways of protecting cell from apoptosis.

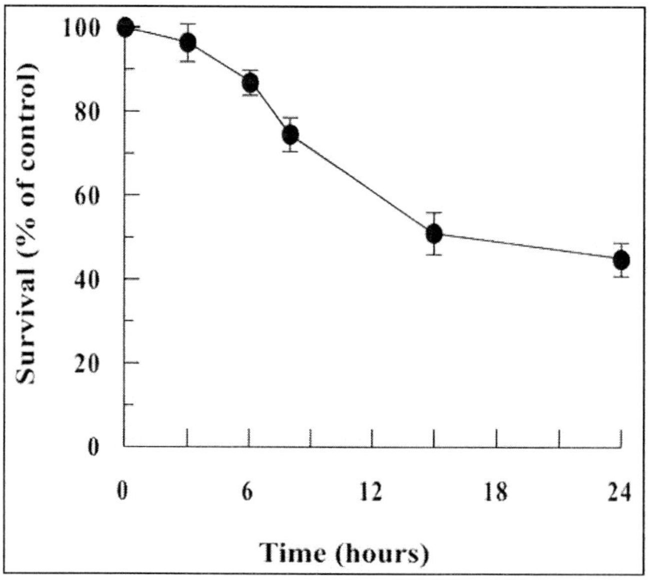


Figure 7. Survival of cerebellar granule cells undergoing apoptosis. Cells were switched to serum-free medium containing low $K^+$ (S-K5) for different times. Cell viability was determined by counting the number of intact nuclei and is expressed as the percentage of control cells to which 100% value was given.

# 3. Apoptosis of Cerebellar Granule Cells. A Survey

In the variety of the parts and processes of the cells somehow involved in cell apoptosis, there in now general consensus on some factors to be absolutely considered in the attempt to describe how cells die via apoptosis. They are reactive oxygen species and the antioxidant system including the superoxide dismutase, catalase and glutathione peroxidase, the proteolitic system including the proteasomes and the caspase family protease, the mitochondrial proteins released during apoptosis, the adenine nucleotide translocator both as responsible for ATP supply to cytosol and as component of the mitochondrial permeability transition pore, the pore itself.

In order, to facilitate the understanding of the apoptosis mechanisms, we will discuss all the experimental findings as shown in the apoptosis time-course in CGCs (Figure 8).

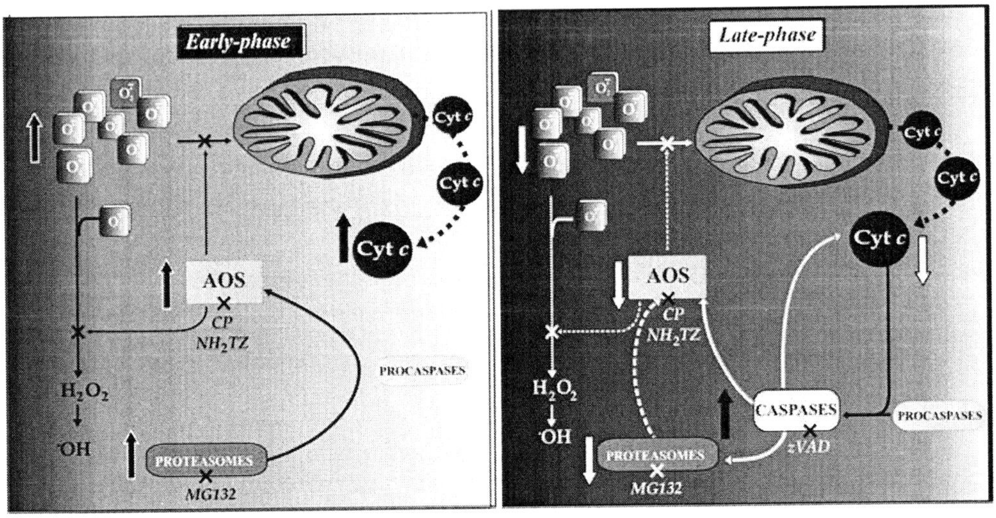


Figure 8. The *EARLY-* and *LATE-phase* of apoptosis: in *EARLY-phase* (0-3 hours) production of ROS occurs, modulation of which is affected by the antioxidant system (AOS), which, in turn, is modulated by proteasomes. Then cytochrome *c* (cyt *c*) is released from the mitochondria. In the *LATE-phase* of apoptosis (3-8 hours) the ROS level reaches the steady-state, there is cyt *c*-dependent caspase activation, the degradation of AOS due to both proteasome and caspase occurs, as does caspase-dependent degradation of the released cyt *c*.

We consider that, following the new mRNA and protein synthesis which occurs upstream ROS production [Schulz *et al.*, 1996], *in vitro* apoptosis of CGCs can occur in two separate phases: an *EARLY-phase* ranging from 0 to 3 hours after the apoptotic stimulus, a *LATE-phase* ranging between 3 and 15 hours. In the *EARLY-phase*, when ROS production increases (see below), the cell components, including antioxidant system and proteasomes are evoked to modulate ROS production for the correct application of the programmed death. The caspases have no role in this phase. In this situation cyt *c* release occurs from still coupled mitochondria [Bobba *et al.*, 2002]. In addition to the activation of the caspase cascade, [Li *et al.*, 1997; Cai *et al.*,1998; Zou *et al.*, 1999; Du *et al.*, 2000], the released cyt *c* work as a ROS scavenger and electron donor the respiratory chain thus generating the electrochemical proton gradient used to drive ATP synthesis [Atlante, 2003b; 2005]. An increase in ATP and a decrease in ADP occur with 1:1 stoichiometry, and with the maximum ATP level found at 3 h apoptosis coinciding with maximum release of cyt *c*. In CGCs cyt *c* release occurs from mitochondria when the permeability transition pore is still close. ROS in the early phase of apoptosis can impair the adeninenucleotide translocator (ANT), decreasing its affinity for ADP. The reduction of ATP supply to cytoplasm is counterbalanced by an increase in the anaerobic glycolysis.

In the *LATE-phase* proteasome activity decreases slowly [Canu *et al.*, 2000; Bobba *et al.*, 2002], but caspase-3 activity increases. This causes a further reduction in AOS enzyme activity and cyt *c* degradation [Bobba *et al.*, 1999]. As a result of this, ROS levels are still high and their action, together with caspase's trigger further ANT impairment, by reducing the transport capability. In this phase, due to caspase action on protein/s different from ANT, ANT become a component of the mPT. Interestingly, in distinction from other systems, in cerebellar granule cell mPTP is dispensable to death [Atlante *et al.*, 2006]. The level of ATP declined up to control levels.

A threshold exists in the pathway leading cells to death with respect to ROS production, Apoptosis, necrosis and another cell death type defined as an "energy catastrophe" have been found to occur by increasing the ROS level [Atlante *et al.*, 2003a].

Each issue summarized above will be reported separately

## 3.1. Oxidative Stress and Cytochrome *c* Release from Mitochondria in Cells Undergoing Apoptosis

### *3.1.1. ROS Levels in Apoptosis*

The ROS levels in cell homeostasis derives from the balance between the ROS production and the ROS removal due to the cell antioxidant system.

*In vivo* and *in vitro* studies have shown a marked increase in ROS levels during apoptosis [Hockenbery *et al.*, 1993; Kane *et al.*, 1993; Greenlund *et al.*, 1995; Rabizadeh *et al.*, 1995]. The role of ROS as intermediates for apoptosis signalling was demonstrated by the following four observations: *(1)* Antioxidants largely prevent some cases of apoptosis [Malorni *et al.*, 1993; Atlante *et al.*, 1998; 2003a]; *(2)* hydrogen peroxide was detected in the cells undergoing apoptosis [Larrick and Wright, 1990]; *(3)* extracellular addition of ROS can induce apoptosis [Lennon *et al.*, 1991] and *(4)* downregulation of intracellular antioxidant level can induce apoptosis [Hockenbery *et al.*, 1990]. Consistently, enzymes of the antioxidant system including manganese superoxide dismutase [Wong, 1995]; glutathione peroxidase [Hockenbery *et al.*, 1993], and copper-zinc superoxide dismutase [Rabizadeh *et al.*, 1995; Greenlund *et al.*, 1995] were found to prevent apoptosis in a variety of systems; glutathione depletion resulted in an increase of ROS levels [Tan *et al.*, 1998].

The possibility that different ROS can cause a different type of death was proposed by Valencia and Morán [2004] who compared the effect of superoxide anion, hydrogen peroxide, and singlet oxygen with respect to the type of cell death; they found that both superoxide anion and singlet oxygen lead to cell death with apoptotic characteristics whereas hydrogen peroxide induces a necrosis-like cell death.

Another factor to be considered is the ROS level/ROS exposure time. As expected oxidative stress has been also implicated in necrotic cell death. Nonetheless, it has been suggested that apoptotic or necrotic cell death can be produced depending on the time of exposure and/or concentration of ROS [Manev *et al.*, 1995; Zhivotowsky *et al.*, 1999; Kruman and Mattson, 1999; Englert and Shacter, 2002; Cole and Perez-Polo, 2002]. Higher concentrations of ROS can induce necrosis, whereas lower concentrations can induce apoptosis in the same cell type [Lennon *et al.*, 1991; Palomba *et al.*, 1996; Eguchi *et al.*, 1997; Leist *et al.*, 1997; Atlante *et al.*, 2003a].

We have investigated the role of ROS level in apoptosis, in particular the dependence of ROS formation on the time after apoptosis induction. In order to assess ROS production, we developed a novel experimental procedure to detect the formation of two reactive oxygen species: superoxide and singlet oxygen, in neuron monolayer primary cultures, thus, making possible the investigation of the effect of certain compounds on reactive oxygen species formation [Atlante *et al.*, 1997; Atlante and Passarella, 1999]. Thus, use was made of two reactive oxygen species detecting systems consisting of ferricytochrome *c* (Fe-cyt *c*) and imidazole-RNO (N, N-dimethyl-4-nitrosoaniline) which allow for the photometric detection of superoxide anion and singlet oxygen, respectively. High superoxide and singlet oxygen production occurs in CGCs undergoing apoptosis (S-K5): their production increases up to 3 hrs and remains rather constant for the following 15 hrs, with no significant variation in S-K25 cells.

Although an increase in ROS production by mitochondria during the early stages of low $K^+$-induced CGC apoptosis has been observed [Cai *et al.*, 1998; Atlante *et al.*, 1998; 2003a], however, the mechanism by which low $K^+$ exposure leads to ROS is a controversial matter [Schulz *et al.*, 1996; Valencia and Moran, 2001; Martin-Romero *et al.*, 2002a,b]. It has been pointed out that most of the ROS burst that potentiates the progress of this apoptosis is localized at the plasma membrane [Valencia and Moran, 2001; Martin-Romero *et al.*, 2002a]. The plasma membrane NADH oxidase activity of neuronal synaptic terminals releases superoxide anion as a co-product and, at maximal stimulation, is the major source of superoxide anion in neurons [Wells and Xu, 1994; May *et al.*, 1995; Martin-Romero *et al.*, 2002b]. Other sources of $O_2^{\cdot -}$ include enzymes such as cytochrome *P*450, in the endoplasmic reticulum, lipoxygenases, cyclooxgenases and xanthine oxidase (XOD). Surprisingly, no xanthine dehydrogenase(XDH)/XOD activity has been detected in neurons [Markley *et al.*, 1973; Dykens *et al.*, 1987], however Atlante *et al.* [1997], taking advantage of a modified XOD assay [Atlante *et al.*, 2000a], showed that XOD activity, derived from XDH via an enzymatic cleavage [McCord, 1985; Dykens, 1994], can produce $O^{\cdot -}_2$ at the beginning of glutamate neurotoxicity in CGCs.

In the nervous system, $O_2^{\cdot -}$ is also produced as a side product of the mitochondrial electron transport chain [Cai and Jones, 1998; Green and Reed, 1998; Atlante *et al.*, 2000b; Herrero and Barja, 2000; Genova *et al.*, 2001]: under physiological conditions, electrons carried by the electron transport chain can leak out of the pathway and pass directly to oxygen, generating $O_2^{\cdot -}$. Mitochondria, the main site of ROS generation in the cell, are also ROS primary target. This, in turn, results in damage to the mitochondrial respiratory chain and, as a consequence, a further increase in ROS generation. A *vicious cycle*, in the terms of Lenaz *et al.* [2000], is thus formed responsible of a number of age-associated dysfunctions of mitochondria and also one of the mechanisms inducing apoptosis.

In the light of the major role of ROS production in apoptosis [Liu *et al.*, 1996; Kluck *et al.*, 1997; Higuchi *et al.*, 1998; Skulachev, 1998; Atlante *et al.*, 1998], the time course of both antioxidant system (AOS, which includes SOD, catalase and the thiol oxidation state) and proteolytic systems (proteasomes and caspases), with respect to their involvement in ROS production and cell survival was investigated. A fast increase in the activity of SOD and catalase and in the thiol oxidation state up to 3 h when cells survive even though ROS production increases. This is consistent with the fact that the steady-state ROS level in CGCs derive from the cell balance between the ROS producing and the ROS-scavenging systems, neither of them fully established to date. Later, in the route leading to death, the antioxidant activity decreases. The same situation applies for proteasome: a slight but consistent increase in activity in the early phase of cell apoptosis, whereas the proteasome activities decline in the *LATE* phase. On the other hand, the time course activation of caspase-3 shows a progressive increase which correlates with apoptosis progression [Armstrong *et al.,* 1997; Eldadah *et al.*, 1997; Atlante *et al.*, 2003a]. The increase in caspase-3 activity strictly correlates with the timing of cyt *c* release, being caspase undetectable in the first 3 hours of apoptosis, when cyt *c* reached its maximum cytosolic level, and already measurable after 4 hours with a maximum reached at 8 hours. Surprisingly, in cell in which AOS is inhibited, no caspase-3 activation occurs, nonetheless cells die. This occurs owing the apoptosis/necrosis shift which takes place when ROS production is high, not regulated by cell AOS. In this regard, in cells in which the antioxidant system is impaired, death occurs via necrosis, as demonstrated by the occurrence of the LDH release, the absence of DNA laddering and the death sensitivity to MK801 which inhibits necrosis without affecting apoptosis [Calissano *et al.*, 1993; Minervini *et al.*, 1997; Atlante *et al.*, 2003a].

The difference in the prevention of ROS decrease found in cells in which proteasome and caspase, separately or together, were blocked with respect to S-K25 cells, suggests that when both proteasome and caspase are active, they participate to the AOS proteolysis in a different manner, being their effects additive. In particular, because AOS activity remains high in cells without proteasome activity in the 3- to 15-h time range, when caspases are active, we conclude that the caspase-dependent AOS degradation requires previous proteasome activity (see below).

### *3.1.2. ROS Targets in Apoptosis*

Although the ROS dependent mechanisms involved in neuronal death are still unclear, it has been shown that ROS can be active on intracellular signalling pathways involved in cell death/survival. In leukemia cells, induction of ROS resulted in inactivation of MEK1/2, ERK1/2, and p70S6K, as well as a dramatic increase in JNK activation and apoptosis [Yu *et al.*, 2003]. ROS formation also induces Akt phosphorylation [Yu *et al.*, 2004]. In other cells, ROS formation induces the activation of FAS receptor in the absence of ligand, leading cells to apoptotic death mediated by caspases, cyt *c* release, and the activation of mitogen p38/JNK [Zhuang *et al.*, 2000; 2001]

Other targets of the ROS are localized in cytoplasm, including phosphofructokinase (PFK), creatine phosphokinase, and lactate dehydrogenase that are susceptible to oxidative inactivation. In this context it should be noted that the activities of several enzymes susceptible to oxidative inactivation, including PFK, are reportedly reduced in post-mortem brain from Alzheimer's disease patients and in gerbil brain after ischemia/reperfusion [Beal *et al.*, 1993]. Nonetheless, under increased ROS production, many oxidative alterations of mitochondrial membrane components may occur, such as lipid peroxidation or protein thiol oxidation, leading to mitochondrial permeabilization and dysfunction [Vercesi *et al.*, 1997; Kowaltowski and Vercesi, 1999; Kowaltowski *et al.*, 2001]. The complete protection against ROS-induced mitochondrial permeabilization conferred by dithiothreitol [Valle *et al.*, 1993; Kowaltowski *et al.*, 1996], is a strong indication that protein thiol groups are important targets for mitochondrial-generated ROS. One other possible membrane protein targets for ROS oxidation is the ADP/ATP carrier, which possesses four cysteine residues [Klingenberg, 1984; Majima *et al.*, 1994]. Consistently this carrier was shown to be impaired in mitochondrial following light irradiation in the presence of photofrin II [Atlante *et al.*, 1989; 1990].

A comparison made between mitochondrial function and cell viability in 3 h apoptosis [Atlante et *al.*, 2003a] shows that in this time when ROS production is maximum, the cells are essentially intact and mitochondria still coupled. On the other hand, the finding that antioxidants as SOD and GSH and reducing agents as DTT, vitamin E and vitamin C are able in preventing mitochondrial failure observed in CGCs exposed to low $K^+$ in the *LATE PHASE*, provides evidence that ROS production in this model will trigger cell apoptosis.

### *Cytochrome c Release*

In apoptosis ROS cause the release of the cyt *c* from the mitochondria. In the *EARLY* phase of low $K^+$-induced apoptosis of CGCs the ROS production triggers cyt *c* release from mitochondria [Atlante *et al.*, 2000b; 2003b]. It is completely prevented if CGCs undergoing apoptosis are preincubated with SOD; nonetheless cyt *c* release occurs when control cells are preincubated with the superoxide production-system, *i.e.* xanthine plus xanthine oxidase [Atlante *et al.*, 1999; 2000b; 2001; 2003b].

The low mitochondrial ROS production which occurs during cyt *c* release, demonstrate that cyt *c* release is mostly dependent on cytosolic ROS production rather than from the mitochondrial impairment that is accompanied by ROS generation. Since cyt *c* release occurred under conditions in which neither glutamate dehydrogenase (GDH) nor, more importantly, adenylate kinase (ADK) were significantly released, it is absolutely excluded that mitochondria damage is the cause of cyt *c* release. On the other hand, the constantly low levels of leaked ADK and GDH in the cytosolic fraction observed after apoptosis induction and the finding that mitochondria, both isolated and when studied in cell homogenate, are coupled and can synthesize ATP under conditions in which cyt *c* is already mostly released, definitely confirms that cyt *c* release is a physiological event occurring with intact mitochondria capable to drive the oxidative phosphorylation (see below) [Atlante *et al.*, 2003b].

In agreement with such a conclusion, confirmed by the lack of mitochondrial swelling, that mitochondrial permeability transition pore is not involved in cyt *c* release from mitochondria during the apoptotic death of cerebellar granule neurons [Wigdal *et al.*, 2002; Atlante *et al.*, 2006]. Consistently, Wigdal *et al.* [2002] demonstrated that cyt *c* release preceded m$\Delta\psi$ loss in CGCs undergoing apoptotic death further excluding that MPTP opening is important for cyt *c* release in these cells. Finally, cyt *c* release proved to be insensitive to permeability transition pore inhibitors [Bossy-Wetzel *et al.*, 1998; Neame *et al.*, 1998; Krohn *et al.*, 1999; Goldstein *et al.*, 2000; Wigdal *et al.*, 2002; Petrosillo *et al.*, 2003].

It should be noted that proteins of the Bcl-2 family have been largely recognized as involved in the release of apoptotic factors, whereas the attribution to caspases of this role deserves some caution. Interestingly, the role of the mitochondrial cardiolipin in the release of mitochondrial proteins is receiving increasing consensus [Ott *et al.*, 2002; Iverson and Orrenius, 2004].

Although the precise mechanisms by which the permeability of the outer mitochondrial membrane (OMM) becomes compromised during apoptosis and by which the apoptogenic factors are released remain elusive, the role of proteasome in the cyt *c* release is already shown: both MG132 and Lactacystin (LC), reversible and irreversible proteasome inhibitors respectively [Figueiredo-Pereira *et al.,* 1994; Fenteany *et al.,* 1995], were found to prevent cyt *c* release in CGCs undergoing apoptosis, thus rescuing cells from apoptosis. At present the proteasome mechanism must remain a matter of speculation. Proteasomes could act on unknown factors involved in cyt *c* release, for instance Bcl2, which is a mitochondrial-localized anti-apoptotic protein candidate, to undergo degradation via proteasomes. Contrarily, we can exclude that cyt *c* release occurs due to caspase action. Such a conclusion derives from z-VAD-fmk, a broad-range inhibitor of caspases, experiments which show that cyt *c* release is not mediated by caspases, since it is not blocked by z-VAD-fmk [Bobba *et al.*, 1999], although this inhibitor largely prevents the apoptotic process [Atlante *et al.*, 2006]. On the other hand, z-VAD-fmk when added at the time when the cytoplasmic cyt *c* level reached its maximum (*i.e.* 3 h) prevented the degradation of the released cyt *c*. This finding indicated that cyt *c* is degraded in a caspase-dependent way, in fact, in opposition to the 0-3h apoptosis situation, in the 3-15h time-range the cyt *c* amount remains constant in the mitochondria, but it strongly decreases in the cytosol; this showing that the caspase dependent proteolysis of cyt *c* occurs [Bobba *et al.*, 1999]. On the other hand, there is general consensus that cyt *c* release and the activity of caspases are strictly correlated. Once the cyt *c* level in cytosol rises above a certain threshold, interaction with other cytoplasmic factors such as Apaf-1 and ATP initiates the activation of caspase-9 [Gallo *et al.*, 1987; Martin and Green, 1995; Li *et al.*, 1997] (see below). It is well known that caspase-3 is the most prominently caspase involved in neural tissue [Cohen, 1997; Armstrong *et al.*, 1997; Moran *et al.*, 1999] and its activation is reported to be a key event in the execution of neuronal apoptosis. Notice that different results have been reported in studies carried out with a different neuronal cell type [Canu *et al.,* 1998].

The picture emerging from the above findings with respect to the cyt *c* release from mitochondria of CGCs is the following: at 3 h maximum cyt c release occurs in a ROS dependent process in which no caspase activity is required [Bobba *et al.*, 1999]; ROS production is regulated by the antioxidant systems modulated in turn by the proteasome. The extra-AOS activity in cells without proteasome activity results in preventing both cyt *c* release and the caspase cascade. Given that cyt *c* release, in cells in which the antioxidant system is impaired, is reduced by preventing proteasome activity, then this shows that cyt *c* release depends also on other protein(s) sensitive to the proteasome activity. The caspase cascade [Budihardjo *et al.*, 1999; Lee and Wei, 2000] contributes to AOS, proteasome and cyt *c* degradation, thus increasing ROS production and leading to programmed cell death.

### 3.1.3. What is the Role Played by the Released Cyt c in the Cytosol?

Apart from both the manner by which cyt *c* is released from mitochondria and how its release is modulated by antioxidant and proteolytic systems, the question then arises as to the functional role played by the cytoplasmic pool of cyt *c* relationship with apoptotic death.

It is commonly accepted that the released cyt *c* play a major role in the assembly of the "apoptosome" a large cytosolic protein complex. Cyt *c* binds Apaf-1 increasing the affinity of Apaf-1 for dATP and inducing a subsequent conformational change that exposes the caspase-recruitment domains of Apaf-1. These domains interact with procaspase-9 to generate a holoenzyme capable of activating caspases-3 and -7. A proteolytic cascade is then initiated with the cleavage of procaspases -2, -6, -8, and -10, leading to the dismantling of the cell [Gallo *et al.,* 1987; Martin and Green, 1995; Hampton *et al.*, 1998. Slee *et al.*, 1999; Van de Craen *et al.*, 1999; Waterhouse *et al.*, 2001]. At this stage, release of cyt *c* and/or at least one other protein from the intermembrane space is recognized as a critical step in apoptosis in all systems in which it has been studied.

Despite the large body of literature dealing with cyt *c* release from mitochondria in cells undergoing apoptosis [Gorman *et al.*, 1996; Kluck *et al.*, 1997; Du *et al.*, 1997; Bossy-Wetzel *et al.*, 1998; Jürgensmejer *et al.*, 1998; Bobba *et al.,* 1999; 2002; 2004; Ott *et al.*, 2002; Scorrano *et al.*, 2002] and its role in caspase activation [Liu *et al.*, 1996; Kluck *et al.*, 1997; Zou *et al.*, 1999], many features of the relationship between apoptosis and cyt *c* release remain obscure at least in part because of contradictory reports. Since apoptotic pathways can differ from each other depending on the type of cells investigated as well as on the manner by which apoptosis is induced, a variety of responses is to be expected.

Nonetheless, cyt *c,* released from CGC mitochondria, essentially coupled and intact, as an intact, functionally active protein, operates as a ROS scavenger and as a respiratory substrate: it is able to cause proton ejection, membrane potential generation and to drive ATP synthesis and export into the extramitochondrial phase. Hence the released cyt *c* can contribute to provision of the ATP required for programmed cell death to occur [Atlante *et al.*, 2000b; 2003b]. Thus, it is able to accept electrons from ROS, to supply electrons to the cytochrome oxidase and to drive the synthesis of ATP [Atlante *et al.*, 2003b]. The direct oxidation of cyt *c* by cytochrome oxidase at mitochondrial contact sites [Ardail *et al.*, 1990] seems unlikely in view of oxidase structure and apparent inability to span the two mitochondrial membranes. Moreover, a structurally distinct type of contact sites, named bridge contact sites, was revealed in brain mitochondria [Perkins *et al.*, 2001], which might play a role in maintaining the structural integrity of the outer and inner membrane systems, as well as in allowing cyt *c* oxidation. In addition, the cyt *c* bound to its antibody (mAb-cyt *c*) is also recognised by mitochondria and oxidised by cytochrome c oxidase. The presence a variety of proteins/factors released from mitochondria during apoptosis [Susin *et al.*, 1999a; Adrain *et al.*, 2001; van Loo *et al.*, 2001; Cande *et al.*, 2002; Lindholm *et al.*, 2004], poses the question as to whether the mitochondrial energization can be due to the combined action of several proteins. In fact, mitochondria contain and release proteins, besides cyt *c*, involved in the apoptotic cascade, such as AIF (apoptosis inducing factor), which seems to induce nuclear condensation [Hu *et al.*, 1998; Patterson *et al.*, 2000; Zamzami and Kroemer 2001] and a protein that promotes certain forms of apoptosis, possibly by neutralising one or more members of the family of anti-apoptotic proteins, i.e. Smac/DIABLO [Du *et al.*, 2000; Adrain *et al.*,

2001; Ekert *et al.*, 2001]. However, because the cytosolic fraction depleted of cyt *c* fails in causing proton ejection and $\Delta\Psi$ generation, and consequently to drive ATP synthesis, no doubt that, at least in CGCs, the pool of cytosolic and mitochondrial released components do not contain, besides cyt *c*, any other reducing acceptor to be used in oxidative phosphorylation [Atlante *et al.*, 2003b].

In conclusion, in addition to its role in apoptosome formation, cyt *c* acts as a bi-functional factor: *i)* it can remove ROS as a scavenger; *ii)* it can drive ATP synthesis, perhaps regulating the apoptosis time course.

These findings are consistent with the elegant Skulachev's paper in which it is suggested that cyt *c* can work as an antioxidant [Pereverzev *et al.*, 2003], by transferring electrons from superoxide anion to molecular oxygen via the respiratory chain. A variety of stimulating considerations concerning the manner by which cyt *c* works are also reported.

If the redox state of cyt *c* is important as part of its role in apoptosis, the protein must have a capacity to relay that information to an effector molecule, presumably Apaf-1 or pro-caspase-9. This subject is still matter of debate. This might be in the form of passing on electrons, or receiving electrons, as the cyt *c* acts in its holoenzyme state [Hampton *et al.*, 1998]. However, Kluck *et al.* [1997] ruled this out by using the substituted forms of cyt *c*, while Pan *et al.* [1999] found that the reduced form is not required, suggesting that there is no requirement for the passing on of electrons. The second mechanism for the relaying of information in signalling pathways is through the use of conformation changes within the protein. Certainly, the structure of cyt *c* is not rigid, and changes have been recorded in cyt *c* on binding to components of the electron transport chain [Hildebrandt *et al.*, 1993] and on the interaction with various surface properties [Tobias *et al.*, 1996]. More importantly here, cyt *c* undergoes conformational changes on reduction and oxidation [Kar *et al.*, 1994; Calver *et al.*, 1997]. Such changes have been suggested to be significant and perhaps modulating the local dipole moments, with a resultant influence on the interactions with other proteins [Kar *et al.*, 1994]. Therefore, if the reduction state of cyt *c* is important for its proapoptotic activity, changes in conformation, which are a result of its redox state might be the key switch in such a mechanism [Hancock *et al.*, 2001].

It should be noted that apparently, $O_2^-$ detoxification is not the only antioxidant function of cyt *c*. Recently Iwahashi *et al.* [2002] reported that cyt *c* catalyses formation of the penthyl and octanoic acid radicals from linoleic acid hydroperoxide. This may be a mechanism of removal of fatty acid hydroperoxides from the outer leaflet of the mitochondrial inner membrane [Goglia and Skulachev, 2003]. Such a process was assumed to complete the removal from the inner membrane of the fatty acid hydroperoxides, initiated by uncoupling protein-mediated electrophoretic pumping of anions of these hydroperoxides from the inner to the outer membrane leaflet (for details see Goglia and Skulachev, 2003).

### 3.2. A Double Role of Mitochondria in CGCs Undergoing Apoptosis

Two general approaches can be used to investigate mitochondrial physiology. In the first, isolated mitochondria are prepared, but the incubation conditions and experimental protocols are designed to imitate as accurately as possible the environment of the intact cell. A temptation is to modify the experimental conditions in order to amplify the phenomenon to be studied.

The second approach is to work with intact cells, where the gain of physiological relevance is balanced by the greater complexity of the preparation and the difficulty in gaining access to the in situ mitochondria. Research in our group has focused on the development of techniques to improve access to the bioenergetics of the in situ mitochondria, with particular relevance to the investigation

In order to investigate the role of mitochondria in bioenergetic apoptosis, the availability of proper experimental systems is strictly necessary. We used *(i)* neurons in suspension or *(ii)* homogenates of neurons or *(iii)* of cultured neurons attached to glass coverslips, both of them containing coupled mitochondria [Atlante *et al.,* 1996; 1998; 2003b] and *(iv)* isolated coupled mitochondria thus allowing investigation of mitochondrial bioenergetics.

Thus, we found that that impairment of cellular respiration takes place as an early event in low [$K^+$]-induced apoptosis of cultured CGCs. This conclusion is supported by the observed reduction in oxygen uptake rate both by cell suspension incubated with glucose and by cell homogenate incubated with succinate. In the latter case, the stimulation of $O_2$ uptake exerted by ADP provides a measure of the ability of mitochondria to synthesize ATP. The very rapid decrease of RCI (see above) value during low [$K^+$] exposure shows that ATP synthesis is rapidly and progressively impaired. Such an impairment of mitochondrial oxidative phosphorylation tightly correlates, on the one side with the extent of apoptosis, and on the other side with the extent of antiapoptotic action of different substances such as IGF-I, actinomicin D and cycloheximide or with ROS scavengers. This, in turn, suggests that cellular respiration impairment is an early step of the cell death program via apoptosis and not a side or late effect as indicated by the finding that 50% reduction of mitochondrial respiration occurs when only 10–15% of all CGCs present in culture exhibit the signs of apoptotic death. Consistently, that mitochondrial dysfunction is not event necessary for progression of apoptosis has been confirmed [Jekabsons and Nicholls, 2006].

### *3.2.1. ATP Production and Level in CGC Apoptosis*

One of the outstanding problems in the role of mitochondria in apoptosis concerns cellular ATP production and utilization. ATP or available energy has been identified as one of the key factors determining whether cell demise occurs via apoptosis or necrosis. A rapid decline in ATP and mitochondrial function is associated with necrosis [Eguchi *et al.*, 1997], whereas apoptosis is characterized by the maintenance of ATP at or near control levels and by late in the loss of mitochondrial function [Ankarcrona *et al.*, 1995; Plaetzer *et al.*, 2002; Atlante *et al.*, 2005]. Thus, when ATP levels are reduced, typical apoptotic stimuli cause necrosis instead of the expected apoptosis [Leist *et al.,* 1997; Nicotera *et al.*, 1998; Atlante *et al.*, 2005]. Conversely, during a toxic insult to neuronal cells, normally expected to produce necrosis, supplementation of cellular energy reserves with D-glucose [Delgado-Esteban *et al.*, 2000] reduces necrosis and shifts the form of cell death towards apoptosis. The requirement of apoptosis for ATP indicates that damage of cellular targets involved in energy supply might play a crucial role in the mode of cell death being executed. An insult that rapidly impairs neuronal ATP reserves may be expected to primarily produce necrotic cell death.

The finding that depletion of ATP during apoptosis prevents the execution of apoptosis, but converts the cell death pathways to necrosis, is consistent with the fact that the apoptotic program has energy-requiring steps, including: *(i)* apoptosome complex formation [Jiang and Wang, 2000; Genini *et al.*, 2000] and processing of pro-caspase-9 [Nicotera *et al.*, 2000]; *(ii)* chromatin condensation and apoptotic body formation [Kass *et al.*, 1996]; *(iii)* phosphorylation of kinases and other pro-apoptotic proteins [Priault *et al.*, 1999; Shohat *et al.*, 2002; Brodie and Blumberg, 2003]; *(iv)* externalization of phosphatidylserine [Gleiss *et al.*, 2002].

Thus, in apoptosis the role both of ATP, mostly produced in the mitochondria by oxidative phosphorylation, and consequently of the mitochondrial adenin-nucleotide translocator (ANT), through which the ATP reaches the cytosol, is crucial. Notice that ANT is also a component of the mPTP.

The level of ATP increases with maximum in 180 min in CGCs *en route* to apoptosis [Giannattasio *et al.*, 2003; Atlante *et al.*, 2005]. In other systems, a general ATP depletion has been shown in apoptosis: for instance ATP depletion was associated with apoptotic cell death in glia cells and neuronal subpopulations as well as in rat embryonal or adult cortical neurons [Tatton *et al.*, 1996; Almeida and Bolanos, 2001; Aito *et al.*, 2002]. That this depends on features of the experimental system used and/or on the type of neuronal cell death insult remains to be established.

The gradual ATP increase results in the increase of the cell energy charge and in the alteration of the steady state levels of the adenine nucleotides towards equilibrium as shown by the increase in the ratio $[ATP][AMP])/[ADP]^2$ up to a value of 1, which is similar to the reported equilibrium constant for the adenylate kinase reaction [Golding *et al.*, 1995]. Notice that the activity of adenylate kinase does not change *en route* to apoptosis. Nonetheless. the increase in the content of ATP in S-K5 cells is accompanied by a stoichiometric decrease in the level of ADP, but by unchanged levels of AMP and its breakdown products including IMP, inosine and adenosine. This strongly suggests that phosphorylation of ADP rather than new synthesis of adenine nucleotides is responsible for the extra content of ATP. It is noteworthy that the maximum level of ATP is reached at 3 h after apoptosis induction that is the time at which maximum cyt *c* release was found from CGCs undergoing apoptosis [Bobba *et al.*, 1999]; since the released cyt *c* can generate the electrochemical proton gradient and drive ATP synthesis in isolated mitochondria *in vitro* [Atlante *et al.*, 2003b], it might be that cyt *c* contributes to ATP extrasynthesis during CGC apoptosis by increasing the electrochemical proton gradient, in addition to that produced in glycolysis.

Particular interest attaches to how CGCs provide ATP in apoptosis. Investigating as to how ATP is synthesized, we show that apoptosis can occur even when the oxidative phosphorylation is completely blocked, being the glycolytic substrate-level phosphorylation the main ATP source.

In S-K5 cells increasing production of L-lactate *en route* to apoptosis occurs [Atlante *et al.*, 2005]: this suggests that glycolysis contributes to ATP synthesis via substrate-level phosphorylation, even if this occurs to a minor extent compared with ATP synthesis by oxidative phosphorylation. Consistently, Atlante *et al.* [2005] showed that complete inhibition of the oxidative phosphorylation, due to oligomycin, resulted in a significant decrease in cellular energy charge; the latter arose from a reduction in ATP and a minor decrease in ADP, but with a significant increase in AMP level. On the other hand, when a very substantial increase in the production of L-lactate occurred, the cells died by apoptosis even in the presence of oligomycin. This strongly suggested that ATP itself is used for apoptosis to occur independently from how it is synthesized. Citrate, which can inhibit largely the substrate-level phosphorylation in glycolysis, proved to inhibit the production of L-lactate, but with a different effectiveness in control and apoptotic cells; the reason for this difference is at present unknown. As oligomycin, citrate alone had no effect on the survival of S-K5 cells as well as on the death type which is still apoptosis. Indeed, the ATP levels measured under these conditions are different, however it might speculate that in the presence of oligomycin ATP is synthesized, as shown by L-lactate production, but it is rapidly used in processes necessary for apoptosis to occur. Thus, it may be conclude that glycolysis can provide the ATP required for apoptosis when oxidative phosphorylation is impaired. The contribution of glycolysis to the synthesis of ATP required for apoptosis could be of particular importance in situations where mitochondrial function is impaired, such as in oxygen deficit in ischemic brain. This is consistent with the increase in the activity of phosphofructokinase as a function of time observed in cell homogenates in which mitochondria coupling begin to decrease when survival is essentially 100% [Atlante *et al.*, 1998]. The facultative role of mitochondrial oxidative phosphorylation in apoptosis is also consistent with reports which show that apoptosis can occur in cells lacking functional mitochondria [Jacobson *et al.*, 1993; Gamen *et al.*, 1995; Comelli *et al.*, 2003], and that anaerobic glycolysis with L-lactate production can perpetuate cell survival when mitochondrial ATP synthesis is suppressed in striated muscle cells [Opie *et al.*, 1992]. Interestingly, it was reported that ATP synthesized in the glycolysis plays a critical role in preventing the collapse of the mitochondrial membrane potential and thus apoptotic cell death [Beltran *et al.*, 2000], suggesting that ATP itself independently on how it is synthesized can support cell viability. Paradoxically, in cell undergoing death the role of mitochondria as ATP producers is secondary with respect to that as cell "arsenal" of death factors.

### 3.2.2. The Role of the Mitochondrial ANT in Apoptosis

ANT is a ~30 kDa protein exclusively localized to the IM, is involved in the specific exchange of ADP and ATP [Fiore *et al.*, 1998]. ANT is an antiporter whose function is strictly required for ATP produced on the matrix side of IM by the $F_{0/1}$ ATP synthase to be exported to the rest to the cell. Genetic inactivation of ANT therefore results into a severe metabolic impairment [Graham *et al.*, 1997]. Transport is driven by the $\Delta\Psi_m$ leading to the net import of one positive charge per reaction cycle (exchange of $ATP^{4-}$ by $ADP^{3-}$). On the other hand, beside export of ATP, ANT also participates in the mitochondrial permeability transition pore (mPTP) formation, together with other proteins including VDAC and Bax-Bcl-2 in the outer membrane and cyclophillin D in the matrix [for refs. see Halestrap *et al.*, 2002; Le Bras *et al.*, 2005]. Although the mPTP is suggested to be an important mediator of apoptosis, the mechanism underlying its opening, particularly in regard to ANT, and its involvement in apoptotic release of mitochondrial proteins, including cyt *c*, is at present the subject of ongoing debate [for refs. see Tatton and Chalmers-Redman 1998; Vieira *et al.*, 2000; Gulbins *et al.*, 2003]. It has been proposed that pro-apoptotic Bcl-2 family members such as Bax and Bak cause opening of the mPTP, whereas anti-apoptotic members, such as Bcl-2, favour closure of this channel [Reed, 1997; Belzacq *et al.*, 2003]. It was shown that mitochondria from liver of ANT knockout mice still possess mPTP activity, suggesting that ANT is a non-essential component of the mPTP and may not be required for mPTP-associated cell death [Kokoszka *et al.*, 2004].

Recently, Atlante *et al.* [2006] have investigated this subject using CGCs as a model system, measuring ADP/ATP exchange and mitochondrial permeability transition (mPT) as a function of time after induction of apoptosis in the absence or presence of compounds designed to selectively affect ROS production and the AOS and proteolytic system. The function of the ANT *en route* to apoptosis is as follows: in early apoptosis a decrease in the transport efficiency of the ANT occurs, caused by a ROS-mediated post-translational modification as shown by its prevention by SOD and resulting in reduced affinity for ADP. The application of experimental conditions described by Atlante *et al.* [2003b], in which either antioxidant or proteolytic systems are impaired with modulation of ROS production, helps to substantiate the above conclusion in that the activity of the AOS, which is higher in the absence of proteasomes, results in reduced impairment of the ANT; in contrast, a deficit in the AOS system and consequent increase in ROS production leads to further ANT damage. It should be noted that impairment of the ANT has been suggested to be due to ROS-dependent cardiolipin oxidation [Nakagawa, 2004b]. However, in rat liver mitochondria we have already found that ROS can impair the ANT carrier as investigated here, in a manner prevented by the thiol reagent mersalyl. This raises the possibility that thiol(s) in ANT molecules may be the target of ROS [for refs. see Atlante *et al.*, 1989; 1990; Le Bras *et al.*, 2005]. This might suggest a chemical modification in the carrier molecule as responsible for the change in ANT activity; unfortunately a similar experiment cannot be carried out in CGCs undergoing apoptosis.

In late apoptosis, together with CsA-sensitive mPTP opening, a further progressive decrease in ANT transport function, *i.e.* of ADP/ATP exchange, which can be prevented by the caspase inhibitor z-VAD, occurs. Consistently Xia *et al.* [2002] found that recombinant caspase-3 directly induced mitochondrial PTP opening in a cyclosporine A-inhibitable manner and the associated mitochondrial apoptotic phenotype. However ANT itself was excluded as the caspase target because the ANT protein did not change in either amount or molecular weight during apoptosis as measured by western blotting.

In the light of the inverse overlapping sensitivity of ANT and mPTP both to the caspase inhibitor and to other inhibitors used to block cell systems participating in apoptosis, and because no change occurs in either the amount or the size of ANT, it is concluded that in CGCs undergoing apoptosis a functional alteration of ANT takes place resulting in ANT becoming a functional component of the pore responsible for the mPT. Such a conclusion is supported by the inverse correlation between the kinetics of loss of transport efficiency by the ANT and the increase in mPTP opening, as calculated both in S-K5 cells and in cells treated with inhibitors at 5–8 h after induction of apoptosis.

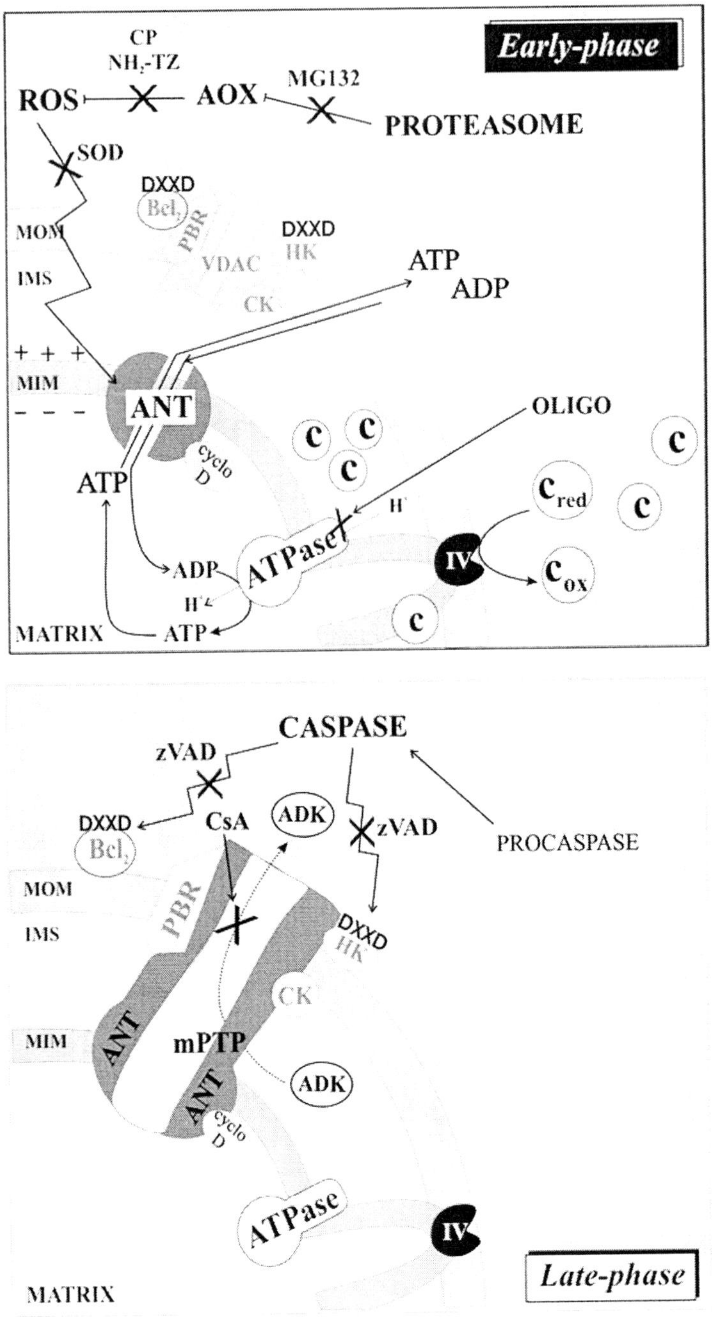


Figure 9. ANT alteration to become a mPTP component. In *EARLY-apoptosis* (0–3 h) a decrease in the transport efficiency of ANT occurs, caused by a ROS-mediated post-translational modification as shown by its prevention by superoxide dismutase (SOD). In *LATE-apoptosis* (3–8 h), together with CsA sensitive-mPTP opening, a further progressive decrease in the ANT transport system, *i.e.* of the ADP/ATP exchange, which can be prevented by z-VAD, the caspase inhibitor, occurs. C, cyt *c*; CK, creatine kinase; $C_{ox}$, oxidized cyt *c*; $C_{red}$, reduced cyt *c*; CsA, cyclosporine A; cyclo D, cyclophillin D; DXXD, caspase consensus sequence; mPTP, mitochondrial Permeability Transition Pore; OLIGO, oligomycin; PBR, peripheraltype benzodiazepine receptor.

A possible target of caspase-dependent proteolysis is Bcl-2 and/or HK, the only mPTP proteins endowed with a caspase consensus sequence. In this regard, it has been shown that Bcl-2 protein is cleaved at Asp-34 by caspases during apoptosis [Kirsch *et al.*, 1999]. The possibility that in mPTP formation caspase action on Bcl-2/HK and the consequent effect on carrier function depend on Bax insertion into the mitochondria [Belzacq *et al.*, 2003; Sharpe *et al.*, 2004; Verrier *et al.*, 2004] cannot be ruled out. Interestingly, ANT residues that can bind Bax have already been identified [Halestrap *et al.*, 2002]. In contrast to Kokoszka *et al.* [2004], Atlante *et al.* [2006] confirmed that mPT occurs as a result of modification of the ANT and that no mPT can take place without this alteration in the ANT that leads to its becoming a pore component. We show that in CGC apoptosis, mPTP is not needed for apoptosis to occur or for the release of cyt *c*, consistent with previous findings [Nakagawa *et al.*, 2005].

Although these results show that the mPTP is not required for the early stages of apoptosis, we cannot rule out that it may be involved in the final stages (up to 24 h), when leaky uncoupled mitochondria finally release their own contents.

In summary, *en route* to apoptosis, a double alteration of the ADP/ATP translocator occurs, owing to the action of ROS in the early phase and of caspase in the late phase. The action of caspase alone can trigger the mPT. This process depends on proteolysis by caspases of mPTP protein component(s) other than the ANT, because no change in either the amount or molecular weight of the ANT takes place during apoptosis. Finally, we show that the mPTP is not required for either cyt *c* release or for apoptosis to occur (Figure 9).

## Conclusion

If the research into apoptosis is at a particularly interesting stage in that many more genes are now known to regulate death, nonetheless how they are activated remains to be established. In particular, the chronological study of apoptosis indicates that early change in mitochondrial bioenergetics take place without a significant cell damage, thus to elucidate how mitochondria modifications participate to the processes leading the cell to death is an exciting challenge for the biochemists. In the good old days, the picture of the mitochondrial universe was simple. Good mitochondria were producing ATP; evil mitochondria were producing ROS. Mitochondrial ROS were, of course, the absolute evil; they augmented diseases, damaged genes, and made us age miserably. These good old days of simple biology are long gone in the wake of new findings, and the emerging concept of mitochondrial ROS production is becoming more and more complex as new elements are discovered. Understanding apoptosis could lead to development of therapeutic pharmacological agents either to decrease or to increase the susceptibility of a variety of cells to apoptosis.

## References

Adrain, C. & Martin, S.J. (2001) The mitochondrial apoptosome: a killer unleashed by the cytochrome seas. *Trends Biochem. Sci., 26*, 390-397.

Adrain, C., Creagh E.M. & Martin S.J. (2001) Apoptosis-associated release of Smac/DIABLO from mitochondria requires active caspases and is blocked by Bcl-2. *EMBO J., 20*, 6627–6636.

Adrain, C., Creagh, E.M. & Martin, S.J. (2002) Caspase Cascades in Apoptosis. Caspases-their role in cell death and cell survival. In: Marek Los and Henning Walczak, Editor. *Moleculare Biology Intelligence Unit 24* (pp. 41-51). New York.

Aito, H., Aalto, K.T. & Raivio, K.O. (2002) Biphasic ATP depletion caused by transient oxidative exposure is associated with apoptotic cell death in rat embryonal cortical neurons, *Pediatr. Res., 52,* 40– 45.

Alavez S., Pedroza D. & Moran J. (2003) Mechanisms of cell death by deprivation of depolarizing conditions during cerebellar granule neurons maturation. *Neurochem. Int., 43,* 581-590.

Almeida, A & Bolanos, J.P. (2001) A transient inhibition of mitochondrial ATP synthesis by nitric oxide synthase activation triggered apoptosis in primary cortical neurons. *J. Neurochem., 77,* 676– 690.

Andreyev, A.Y., Kushnareva, Y.E. & Starkov, A.A. (2005) Mitochondrial metabolism of reactive oxygen species. *Biochemistry, 70,* 200-214

Ankarcrona, M., Dypbukt, J.M., Brune, B. & Nicotera, P. (1994) Interleukin-1 beta-induced nitric oxide production activates apoptosis in pancreatic RINm5F cells. *Exp. Cell. Res., 213*, 172-177.

Ankarcrona, M., Dypbukt, J.M., Bonfoco, E., Zhivotovsky, B., Orrenius, S., Lipton, S.A. & Nicotera, P. (1995) Glutamate-induced neuronal death: a succession of necrosis or apoptosis depending on mitochondrial function. *Neuron, 15*, 961–973.

Antonsson, B., Conti, F., Ciavatta, A., Montessuit, S., Lewis, S., Martinou, I., Bernasconi, L., Bernard, A., Mermod, J.J., Mazzei, G., Maundrell, K., Gambale, F., Sadoul, R. & Martinou J.C. (1997) Inhibition of Bax channel-forming activity by Bcl-2. *Science, 277,* 370–372.

Arceci, R.J. (1996) Tumor cell survival and resistance to therapy. *Curr. Opin. Hematol., 3,* 279-287.

Ardail, D., Privat, J.-P., Egret-Charlier, M., Levrat, C., Lerme, F. & Louisott, P. (1990) Mitochondrial contact sites. Lipid composition and dynamics. *J. Biol. Chem., 265,* 18797-18802.

Armstrong, R.C., Aja, T.J., Hoang, K.D., Gaur, S., Bai, X., Alnemri, E.S., Litwack, G., Karanewsky, D.S., Fritz, L.C. and Tomaselli, K.J. (1997) Activation of the CED3/ICE-related protease CPP32 in cerebellar granule neurons undergoing apoptosis but not necrosis. *J. Neurosci., 17,* 553-562.

Atlante, A., Passarella, S., Quagliariello, E., Moreno, G. & Salet, C. (1989) Haematoporphyrin derivative (Photofrin II) photosensitization of isolated mitochondria: inhibition of ADP/ATP translocator. *J. Photochem. Photobiol. B., 4,* 35-46.

Atlante, A., Passarella, S., Quagliariello, E., Moreno, G. & Salet, C. (1990) Carrier thiols are targets of Photofrin II photosensitization of isolated rat liver mitochondria. *J. Photochem. Photobiol. B., 7,* 21-32.

Atlante, A., Gagliardi, S., Minervini, G.M., Marra, E., Passarella, S. & Calissano, P. (1996) Rapid uncoupling of oxidative phosphorylation accompanies glutamate toxicity in rat cerebellar granule cells. *Neuroreport, 7,* 2519–2523.

Atlante, A., Gagliardi, S., Minervini, G.M., Ciotti, M.T., Marra, E. & Calissano, P. (1997) Glutamate neurotoxicity in rat cerebellar granule cells: a major role for xanthine oxidase in oxygen radical formation. *J. Neurochem., 68,* 2038–2045.

Atlante, A., Gagliardi, S., Marra, E. & Calissano, P. (1998) Neuronal apoptosis in rats is accompanied by rapid impairment of cellular respiration and is prevented by scavengers of reactive oxygen species. *Neurosci. Lett., 245,* 127–130.

Atlante, A., Gagliardi, S., Marra, E., Calissano, P. & Passarella, S. (1999) Glutamate neurotoxicity in rat cerebellar granule cells involves cytochrome c release from mitochondria and mitochondrial shuttle impairment. *J. Neurochem., 73,* 237–246.

Atlante, A. & Passarella, S. (1999) Detection of reactive oxygen species in primary cultures of cerebellar granule cells. *Brain Res. Protocols 4,* 266–270.

Atlante, A., Valenti, D., Gagliardi, S. & Passarella, S. (2000a) A sensitive method to assay the xanthine oxidase activity in primary cultures of cerebellar granule cells. *Brain Res Protoc., 6,* 1-5.

Atlante, A., Calissano, P., Bobba, A., Azzariti, A., Marra, E. & Passarella S. (2000b) Cytochrome *c* is released from mitochondria in a reactive oxygen species (ROS)-dependent fashion and can operate as a ROS scavenger and as a respiratory substrate in cerebellar neurons undergoing excitotoxic death. *J. Biol. Chem., 275,* 37159-37166.

Atlante, A., Calissano, P., Bobba, A., Giannattasio, S., Marra, E. & Passarella, S. (2001) Glutamate neurotoxicity, oxidative stress and mitochondria. *FEBS Lett., 497,* 1–5.

Atlante, A., Bobba, A., Calissano, P., Passarella, S. & Marra, E. (2003a) The apoptosis/necrosis transition in cerebellar granule cells depends on the mutual relationship of the antioxidant and the proteolytic systems which regulate ROS production and cytochrome c release en route to death, *J. Neurochem., 84,* 960–971.

Atlante, A., de Bari, L., Bobba, A., Marra, E., Calissano, P. & Passarella, S. (2003b) Cytochrome *c*, released from cerebellar granule cells undergoing apoptosis or excytotoxic death, can generate protonmotive force and drive ATP synthesis in isolated mitochondria. *J. Neurochem., 86,* 591-604.

Atlante, A., Giannattasio, S., Bobba, A., Gagliardi, S., Petragallo, V., Calissano, P., Marra, E. & Passarella, S. (2005) An increase in the ATP levels occurs in cerebellar granule cells en route to apoptosis in which ATP derives from both oxidative phosphorylation and anaerobic glycolysis. *Biochim. Biophys. Acta, 1708,* 50–62.

Atlante, A., Bobba, A., de Bari, L., Fontana, F., Calissano, P., Marra, E. & Passarella, S. (2006) Caspase-dependent alteration of the ADP/ATP translocator triggers the mitochondrial permeability transition which is not required for the low-potassium-dependent apoptosis of cerebellar granule cells. *J. Neurochem., 97,* 1166-1181.

Balàsz, R., Resink, A., Hack, N., Van der Walk, J.B.F., Kumar, K.N., Michaelis, E. (1992) NMDA treatment and $K^+$-induced depolarization selectively promote the expression of a NMDA-preferring class of the ionotropic glutamate receptor in cerebellar granule cells. *Neurosci. Lett., 137,* 109-113.

Barile, M., Brizio, C., Valenti, D., De Virgilio, C. & Passarella, S. (2000) The riboflavin/FAD cycle in rat liver mitochondria. *Eur. J. Biochem., 267,* 4888-4900.

Basanez, G., Nechushtan, A., Drozhinin, O., Chanturiya, A., Choe, E., Tutt, S., Wood, K.A., Hsu, Y., Zimmerberg, J. & Youle, R.J. (1999) Bax, but not Bcl-xL, decreases the lifetime of planar phospholipid bilayer membranes at subnanomolar concentrations. *Proc. Natl. Acad. Sci. U. S. A., 96,* 5492–5497.

Beal, M.F., Hyman, B.T. & Koroshetz, W. (1993) Do defects in mitochondrial energy metabolism underlie the pathology of neurodegenerative diseases? *Trends Neurosci., 16,* 125-131.

Beltran, B., Mathur, A., Duchen, M.R., Erusalimsky, J.D. & Moncada, S. (2000) The effect of nitric oxide on cell respiration: a key to understanding its role in cell survival or death. *Proc. Natl. Acad. Sci. U.S.A., 97,* 14602–14607.

Belzacq, A.S., Vieira, H.L., Verrier, F., Vandecasteele, G., Cohen, I., Prevost, M.C., Larquet, E., Pariselli, F., Petit, P.X., Kahn, A., Rizzuto, R., Brenner, C., Kroemer, G. (2003) Bcl-2 and Bax modulate adenine nucleotide translocase activity. *Cancer Res., 63,* 541–546.

Bobba, A., Atlante, A., Giannattasio, S., Sgaramella, G., Calissano, P. & Marra, E. (1999) Early release and subsequent caspase-mediated degradation of cytochrome c in apoptotic cerebellar granule cells. *FEBS Lett., 457*, 126–130.

Bobba, A., Canu, N., Atlante, A., Petragallo, V., Calissano, P. & Marra, E. (2002) Proteasome inhibitors prevent cytochrome *c* release during apoptosis but not in excitotoxic death of cerebellar granule neurons. *FEBS Lett., 515:* 8-12.

Bobba, A., Atlante, A., de Bari, L., Passarella, S. & Marra, E. (2004) Apoptosis and cytochrome *c* release in cerebellar granule cells. *In Vivo, 18*, 335–344.

Borner, C. & Monney, L. (1999) Apoptosis without caspases: an inefficient molecular guillotine? *Cell. Death Diff., 6,* 497- 507.

Bossy-Wetzel, E., Newmeyer, D.D. & Green, D.R. (1998) Mitochondrial cytochrome c release in apoptosis occurs upstream of DEVD-specific caspase activation and independently of mitochondrial transmembrane depolarization. *EMBO J., 17,* 37– 49.

Bratton, S.B., Walker, G., Srinivasula, S.M., Sun, X.M., Butterworth, M., Alnemri, E.S. & Cohen, G.M. (2001) Recruitment, activation and retention of caspases-9 and -3 by Apaf-1 apoptosome and associated XIAP complexes. *EMBO J., 20*, 998-1009.

Breckenridge, D.G. & Xue, D. (2004) Regulation of mitochondrial membrane permeabilization by BCL-2 family proteins and caspases. *Curr. Opin. Cell Biol., 16*, 647-652.

Brodie, C. & Blumberg, P.M. (2003) Regulation of cell apoptosis by protein kinase c delta. *Apoptosis, 8,* 19– 27.

Budihardjo, I., Oliver, H., Lutter, M., Luo, X. & Wang, X. (1999) Biochemical pathways of caspase activation during apoptosis. *Annu. Rev. Cell. Dev. Biol., 15*, 269–290.

Cai, J. & Jones, D.P. (1998) Superoxide in apoptosis. Mitochondrial generation triggered by cytochrome c loss. *J. Biol. Chem., 273*, 11401–11404.

Cai, J., Yang, J. & Jones, D.P. (1998) Mitochondrial control of apoptosis: the role of cytochrome *c*. *Biochim. Biophys. Acta, 1366,* 139-149.

Cain, K., Brown, D.G., Langlais, C. & Cohen, G.M. (1999) Caspase activation involves the formation of the apoptosome, a large (approximately 700 kDa) caspase-activating complex. *J. Biol. Chem., 274*, 22686–22692

Calissano, P., Ciotti, M.T., Battistini, L., Zona, C., Angelici, A., Merlo, D. & Mercanti, D. (1993) Recombinant human insulin-like growth factor I exerts a trophic action and confers glutamate sensitivity on glutamate-resistant cerebellar granule cells. *Proc. Natl Acad. Sci. USA, 90*, 8752–8756.

Calver, J.F., Hill, J.L. & Dong, A. (1997) Redox-dependent conformational changes are common structural features of cytochrome *c*from various species. *Arch. Biochem. Biophys., 346,* 287–293.

Cande, C., Cohen, I., Daugas, E., Ravagnan, L., Larochette, N., Zamzami, N. & Kroemer, G. (2002) Apoptosis-inducing factor (AIF): a novel caspase-independent death effector released from mitochondria. *Biochimie, 84,* 215-222.

Canu, N., Dus, L., Barbato, C., Ciotti, M.T., Brancolini, C., Rinaldi, A.M., Novak, M., Cattaneo, A., Bradbury, A. and Calissano, P. (1998) Tau cleavage and dephosphorylation in cerebellar granule neurons undergoing apoptosis. *J. Neurosci., 18,* 7061-7074.

Canu, N., Barbato, C., Ciotti, M.T., Serafino, A., Dus, L. & Calissano, P. (2000) Proteasome involvement and accumulation of ubiquitinated proteins in cerebellar granule neurons undergoing apoptosis. *J. Neurosci., 20,* 589-599.

Canu, N. &, Calissano, P. (2003) In vitro cultured neurons for molecular studies correlating apoptosis with events related to Alzheimer disease. *Cerebellum, 2,* 270-278.

Cheng, E.H., Kirsch, D.G., Clem, R.J., Ravi, R., Kastan, M.B., Bedi, A., Ueno, K. & Hardwick, J.M. (1997) Conversion of Bcl-2 to a Bax-like death effector by caspases. *Science, 278,* 1966–1968.

Cohen, G.M. (1997) Caspases: the executioners of apoptosis. *Biochem. J., 326,* 1-16.

Cole, K.K. & Perez-Polo, J.R. (2002) Poly(ADP-ribose) polymerase inhibition prevents both apoptotic-like delayed neuronal death and necrosis after $H_2O_2$ injury. *J. Neurochem., 82,* 19-29.

Comelli, M., Di Pancrazio, F. & Mavelli, I. (2003) Apoptosis is induced by decline of mitochondrial ATP synthesis in erythroleukemia cells. *Free Radic. Biol. Med., 34*, 1190–1199.

Contestabile, A. (2002) Cerebellar granule cells as a model to study mechanisms of neuronal apoptosis or survival *in vivo* and *in vitro*. *Cerebellum, 1*, 41-55. Review

Crompton, M. (1999) The mitochondrial permeability transition pore and its role in cell death. *Biochem. J., 341*, 233–249.

Cull-Candy, S.G., Howe, J.R. & Ogden, D.C. (1988) Noise and single channels activated by excitatory amino acids in rat cerebellar granule neurones. *J. Physiol., 400,* 189-222.

Darmon, A.J., Ley, T.J., Nicholson, D.W. & Bleackley, R.C. (1996) Cleavage of CPP32 by granzyme B represents a critical role for granzyme B in the induction of target cell DNA fragmentation. *J. Biol. Chem., 271*, 21709–21712.

Deigner, H.P., Haberkorn, U. & Kinscherf, R. (2000) Apoptosis modulators in the therapy of neurodegenerative diseases. *Expert Opin Investig Drugs., 9,* 747-764.

Delgado-Esteban, M., Almeida, A. & Bolanos, J.P. (2000) D-Glucose prevents glutathione oxidation and mitochondrial damage after glutamate receptor stimulation in rat cortical primary neurons. *J. Neurochem., 75,* 1618-1624.

De Martino, G.N. & Slaughter, C.A. (1999) The proteasome, a novel protease regulated by multiple mechanisms. *J. Biol. Chem., 274,* 22123-22126.

Desagher, S. & Martinou, J.C. (2000) Mitochondria as the central control point of apoptosis. *Trends Cell. Biol., 10,* 369–377.

Didier, M., Roux, P., Piechaczyk, M., Mangeat, P., Devilliers, G., Bockaert, J. &, Pin, J.P. (1992) Long-term expression of the c-fos protein during the in vitro differentiation of cerebellar granule cells induced by potassium and NMDA. *Mol. Brain Res., 12,* 249-258.

Didier, M., Roux, P., Piechaczyk, M., Verrier, B., Bockaert, J. & Pin, J.P. (1989) Cerebellar granule cell survival and maturation induced by $K^+$ and NMDA correlate with c-fos protooncogene expression. *Neurosci. Lett., 107,* 55-62.

D'Mello, S.R., Galli, C., Ciotti, T. & Calissano, P. (1993) Induction of apoptosis in cerebellar granule neurons by low potassium: inhibition of death by insulin-like growth factor I and cAMP. *Proc Natl Acad Sci U S A., 90,* 10989-10993.

Du, C., Fang, M., Li, Y., Li, L. & Wang, X. (2000) Smac, a mitochondrial protein that promotes cytochrome c-dependent caspase activation by eliminating IAP inhibition. *Cell, 102,* 33–42.

Du, Y., Dodel, R.C., Bales, K.R., Jemmerson, R., Hamilton-Byrd, E. & Paul, S.M. (1997) Involvement of a caspase-3-like cysteine protease in 1-methyl-4-phenylpyridinium-mediated apoptosis of cultured cerebellar granule neurons. *J. Neurochem., 69,* 1382-1388.

Dykens, J.A., Stern, A. and Trenkner, E. (1987) Mechanism of kainate toxicity to cerebellar neurons in vitro is analogous to reperfusion tissue injury. *J. Neurochem., 49,* 1222-1228.

Dykens, J.A. (1994) Isolated cerebral and cerebellar mitochondria produce free radicals when exposed to elevated CA2+ and $Na^+$: implications for neurodegeneration. *J. Neurochem., 63,* 584-591.

Eguchi, Y., Shimizu, S. & Tsujimoto, Y. (1997) Intracellular ATP levels determine cell death fate by apoptosis or necrosis. *Cancer Res., 57,* 1835– 1840.

Ekert, P.G., Silke, J., Hawkins, C.J., Verhagen, A.M. & Vaux, D.L. (2001) DIABLO promotes apoptosis by removing MIHA/XIAP from processed caspase 9. *J. Cell. Biol., 152,* 483–490.

Eldadah, B.A., Yakovlev, A.G. & Faden, A.I. (1997) The role of CED-3-related cysteine proteases in apoptosis of cerebellar granule cells. *J. Neurosci., 17,* 6105-6113.

Englert, R.P. & Shacter, E. (2002) Distinct modes of cell death induced by different reactive oxygen species: amino acyl chloramines mediate hypochlorous acid-induced apoptosis. *J. Biol. Chem., 277,* 20518-20526.

Evan, G. & and Littlewood, T. (1998) A matter of life and cell death. *Science, 281,* 1317-1322.

Fenteany, G., Standaert, R.F., Lane, W.S., Choi, S., Corey, E.J. & Schreiber, S.L. (1995) Inhibition of proteasome activities and subunit-specific amino-terminal threonine modification by lactacystin. *Science, 268,* 726–731.

Figueiredo-Pereira, M.E., Berg, K.A. & Wilk, S. (1994) A new inhibitor of the chymotrypsin-like activity of the multicatalytic proteinase complex (20S proteasome) induces accumulation of ubiquitinprotein conjugates in a neuronal cell. *J. Neurochem., 63*, 1578–1581.

Fiore, C., Tezeguet, V., LeSaux, A., Roux, P., Schwimmer, C., Dianoux, A.C., Noel, F., Lauquin, G.J.M., Brandolin, G. and Vignais, P.V. (1998) The mitochondrial ADP/ATP carrier: structural, physiological and pathological aspects. *Biochimie, 80*, 137–150.

Fleury, C., Mignotte, B. & Vayssiere, J.L. (2002) Mitochondrial reactive oxygen species in cell death signaling. *Biochimie, 84*, 131-141. Review.

Franklin, J.L. & Johnson, E.M. (1992) Suppression of programmed neuronal death by sustained elevation of cytoplasmic calcium. *Trends Neurosci., 15,* 501-508.

Fujimura, M., Morita-Fujimura, Y., Kawase, M., Copin, J.-C., Calagui, B., Epstein, C.J. & Chan, P.H. (1999) Manganese superoxide dismutase mediates the early release of mitochondrial cytochrome *c* and subsequent DNA fragmentation after permanent focal cerebral ischemia in mice. *J. Neurosci., 19,* 3414-3422.

Fujimura, M., Morita-Fujimura, Y., Noshita, N., Sugawara, T., Kawase, M. & Chan, PH. (2000) The cytosolic antioxidant copper/zinc-superoxide dismutase prevents the early release of mitochondrial cytochrome *c* in ischemic brain after transient focal cerebral ischemia in mice. *J. Neurosci., 20,* 2817-2824.

Gallo, V., Ciotti, M.T., Coletti, A., Aloisi, F. & Levi, G. (1982) Selective release of glutamate from cerebellar granule cells differentiating in culture. *Proc. Natl Acad. Sci. U.S.A., 79,* 7919-7923.

Gallo, V., Kingsbury, A., Balazs, R. & Jorgensen, O.S. (1987) The role of depolarization in the survival and differentiation of cerebellar granule cells in culture. *J. Neurosci., 7,* 2203-2213.

Gamen, S., Anel, A., Montoya, J., Marzo, I., Pineiro, A. & Naval, J. (1995) mtDNA-depleted U937 cells are sensitive to TNF and Fas-mediated cytotoxicity. *FEBS Lett., 376*, 15–18.

Genini, D., Budihardjo, I., Plunkett, W., Wang, X., Carrera, C.J., Cottam, H.B., Carson, D.A. & Leoni, L.M. (2000) Nucleotide requirements for the in vitro activation of the apoptosis protein activating factor-1-mediated caspase pathway. *J. Biol. Chem., 275*, 29– 34.

Genova, M.L., Ventura, B., Giuliano, G., Bovina, C., Formiggini, G., Parenti Castelli, G. & Lenaz, G. (2001) The site of production of superoxide radical in mitochondrial Complex I is not a bound ubisemiquinone but presumably iron–sulfur cluster N2. *FEBS Lett., 505,* 364-368.

Ghosh, A., Carnahan, J. & Greenberg, M.E. (1994) Requirement for BDNF in activity-dependent survival of cortical neurons. *Science, 263*, 1618-1622.

Giannattasio, S., Gagliardi, S., Samaja, M. & Marra, E. (2003) Simultaneous determination of purine nucleotides, their metabolites and betanicotinamide adenine dinucleotide in cerebellar granule cells by ionpair high performance liquid chromatography. *Brain Res. Protoc., 10*, 168–174.

Gleiss, B., Gogvadze, V., Orrenius, S. & Fadeel, B. (2002) Fas-triggered phosphatidylserine exposure is modulated by intracellular ATP. *FEBS Lett., 519*, 153– 158.

Goglia, F. & Skulachev, V.P. (2003) A function for novel uncoupling proteins: antioxidant defense of mitochondrial matrix by translocating fatty acid peroxides from the inner to the outer membrane leaflet. *FASEB J., 17*, 1585–1591.

Golding, E.M., Teague Jr, W.E. & Dobson, G.P. (1995) Adjustment of K' to varying pH and pMg for the creatine kinase, adenylate kinase and ATP hydrolysis equilibria permitting quantitative bioenergetic assessment. *J. Exp. Biol., 198*, 1775– 1782.

Goldstein, J.C., Waterhouse, N.J., Juin, P., Evan, G.I. & Green, D.R. (2000) The coordinate release of cytochrome c during apoptosis is rapid, complete and kinetically invariant. *Nat. Cell. Biol., 2*, 156–162.

Gorman, A.M., McGowan, A., O'Neill, C. & Cotter, T. (1996) Oxidative stress and apoptosis in neurodegeneration. *J. Neurol. Sci., 139*, 45-52.

Graham, B.H., Waymire, K.G., Cottrell, B., Trounce, J.A., MacGregor, G.R. & Wallace D.C. (1997) A mouse model for mitochondrial myopathy and cardiomyopathy resulting from a deficiency in the heart/muscle isoform of the adenine nucleotide translocator. *Nature Genet., 16,* 226-234.

Green, D.R. & Reed, J.C. (1998) Mitochondria and apoptosis. *Science, 281,* 1309-1312.

Greenlund, L.J.S., Deckwerth, T.L. & Johnson, E.M. Jr (1995) Superoxide dismutase delays neuronal apoptosis: a role for reactive oxygen species in programmed neuronal death. *Neuron, 14*, 303–315.

Grimm, L.M., Goldberg, A.L., Poirier, G.G., Schwartz, L.M. & Osborne, B.A. (1996) Proteasomes play an essential role in thymocyte apoptosis. *EMBO J., 15,* 3835-3844.

Gross, A., McDonnell, J.M. & Korsmeyer, S.J. (1999) BCL-2 family members and the mitochondria in apoptosis. *Genes Dev., 13,* 1899-1911.

Guchelaar, H.J., Vermes, A., Vermes, I. & Haanen, C. (1997) Apoptosis: molecular mechanisms and implications for cancer chemotherapy. *Pharm. World Sci., 19,* 119-125.

Gulbins, E., Dreschers, S. & Bock, J. (2003) Role of mitochondria in apoptosis. *J. Exp. Physiol., 88*, 85–90.

Haeberlein, S.L. (2004) Mitochondrial function in apoptotic neuronal cell death. *Neurochem. Res., 29,* 521-530.

Halestrap, A.P., McStay, G.P. & Clarke, S.J.B. (2002) The permeability transition pore complex: another view. *Biochimie, 84*, 153–166.

Hampton. MB, Zhivotovsky B, Slater AF, Burgess DH & Orrenius S (1998) Importance of the redox state of cytochrome c during caspase activation in cytosolic extracts. *Biochem. J., 329*, 95–99.

Hancock, J.T., Desikan, R. & Neill, S.J. (2001) Does the redox status of cytochrome *C* act as a fail-safe mechanism in the regulation of programmed cell death? *Free Radic Biol Med., 31,* 697-703. Review.

Harman, D. (1992) Free radical theory of aging. *Mutat. Res., 275*, 257-266.

Harvey, N.L., Trapani, J.A., Fernandes-Alnemri, T., Litwack, G., Alnemri, E.S. & Kumar, S (1996) Processing of the Nedd2 precursor by ICE-like proteases and granzyme B. *Genes Cells. 1,* 673-685.

Herrero, A. & Barja, G. (2000) Localization of the site of oxygen radical generation inside the complex I of heart and nonsynaptic brain mammalian mitochondria. *J. Bioenerg. Biomembr., 32*, 609-615.

Higuchi, M., Honda, T., Prose, R.J. & Yeh, E.T.H. (1998) Regulation of reactive oxygen species-induced apoptosis and necrosis by caspase 3-like proteases. *Oncogene, 17*, 2753–2760.

Hildebrandt, P., Vanhecke, F., Buse, G., Tewfik, S., & Grant, M. (1993) Resonance Raman study of the interactions between cytochrome *c* variants and cytochrome *c* oxidase. *Biochemistry, 32*, 10912–10922.

Hirsch, T., Marchetti, P., Susin, S.A., Dallaporta, B., Zamzami, N., Marzo, I., Geuskens, M., & Kroemer, G. (1997) The apoptosis-necrosis paradox. Apoptogenic proteases activated after mitochondrial permeability transition determine the mode of cell death. *Oncogene, 15,* 1573-1581.

Hockberger, P.E., Tseng, H.Y. & Connor, J.A. (1987) Immunocytochemical and electrophysiological differentiation of rat cerebellar granule cells in explant cultures. *J. Neurosci., 7,* 1370-1383.

Hockenbery, D., Nunez, G., Milliman, C., Schreiber, R.D. & Korsemeyer, S.G. (1990) Bcl-2 is an inner mitochondrial membrane protein that blocks programmed cell death. *Nature, 348,* 334–336.

Hockenbery, D.M., Oltvai, Z.N., Yin, X.M., Milliman, C.L. & Korsmeyer, S.J. (1993) Bcl-2 functions in an antioxidant pathway to prevent apoptosis. *Cell, 75*, 241-251.

Hu, Y., Benedict, M.A., Wu, D., Inohara, N. & Nunez, G. (1998) Bcl-XL interacts with Apaf-1 and inhibits Apaf-1-dependent caspase-9 activation. *Proc. Natl. Acad. Sci. USA, 95*, 4386–4391.

Iverson, S.L. & Orrenius, S. (2004) The cardiolipin-cytochrome c interaction and the mitochondrial regulation of apoptosis. *Arch Biochem Biophys., 423*, 37-46. Review

Iwahashi, H., Nishizaki, K. & Takagi, I. (2002) Cytochrome c catalyses the formation of pentyl radical and octanoic acid radical from linoleic acid hydroperoxide. *Biochem. J. 361*, 57–66.

Jaattela, M., Wissing, D., Kokholm, K., Kallunki, T. & Egeblad, M. (1998) Hsp70 exerts its anti-apoptotic function downstream of caspase-3-like proteases. *EMBO J., 17,* 6124-6134.

Jacobson, M.D., Burne, J.F., King, M.P., Miyashita, T.J., Reed, C. & Raff, M.C. (1993) Bcl-2 blocks apoptosis in cells lacking mitochondrial DNA. *Nature, 361,* 365–369.

Jekabsons, M.B. & Nicholls, D.G. (2006) Bioenergetic analysis of cerebellar granule neurons undergoing apoptosis by potassium/serum deprivation. *Cell Death Differ.* doi: 10.1038/sj.cdd.4401851

Jiang, X. & Wang, X. (2000) Cytochrome c promotes caspase-9 activation by inducing nucleotide binding to Apaf-1. *J. Biol. Chem. 275*, 31199– 31203.

Jürgensmejer, J.M., Xie, Z., Deveraux, Q., Ellerby, L., Bredesen, D. & Reed, C. (1998) Bax directly induces release of cytochrome c from isolated mitochondria. *Proc. Natl. Acad. Sci. U.S.A., 95,* 4997- 5002.

Kahl, R., Kampkotter, A., Watjen, W. & Chovolou, Y. (2004) Antioxidant enzymes and apoptosis. *Drug Metab. Rev. 36*, 747-762.

Kane, D.J., Sarafian, T.A., Anton, R., Hahn, H., Gralla, E.B., Valentine, J.S., Ord, T. & Bredesen, D.E. (1993) Bcl-2 inhibition of neural death: decreased generation of reactive oxygen species. *Science, 262*, 1274-1277

Kar, L., Sherman, S. & Johnson, M.E. (1994) Comparison of protein structures in solution using local conformations derived from NMR data: application to cytochrome *c*. *J. Biomol. Struct. Dyn.*, *12*, 527–558.

Kass, G.E., Eriksson, J.E., Weis, M., Orrenius, S. & Chow, S.C. (1996) Chromatin condensation during apoptosis requires ATP. *Biochem. J.*, *318*, 749– 752.

Kerr, J.F., Wyllie, A.H. & Currie, A.R. (1972) Apoptosis: a basic biological phenomenon with wide-ranging implications in tissue kinetics. *Br. J. Cancer*, *26*, 239-257.

Kingsbury, A.E., Gallo, V., Woodhams, P.L. & Balazs, R. (1985) Survival, morphology and adhesion properties of cerebellar interneurones cultured in chemically defined and serum-supplemented medium. *Brain Res.*, *349*, 17-25.

Kirsch, D.G., Doseff, A., Chau, B.N., Lim, D.S., de Souza-Pinto, N.C., Hansford, R., Kastan, M.B., Lazebnik, Y.A. & Hardwick, J.M. (1999) Caspase-3-dependent cleavage of Bcl-2 promotes release of cytochrome c. *J. Biol. Chem.*, *274*, 21155-21161.

Klingenberg, M. (1984) Molecular aspects of the adenine nucleotide carrier from mitochondria. *Arch. Biochem. Biophys.*, *270*, 1-14.

Kluck, R.M., Bossy-Wetzel, E., Green, D.R. and Newmeyer, D.D. (1997) The release of cytochrome *c* from mitochondria: a primary site for Bcl-2 regulation of apoptosis. *Science*, *275*, 1132-1136.

Kokoszka, J.E., Waymire, K.G., Levy, S.E., Sligh, J.E., Cai, J., Jones, D.P., MacGregor, G.R. & Wallace, D.C. (2004) The ADP/ATP translocator is not essential for the mitochondrial permeability transition pore. *Nature*, *427*, 461–465.

Kowaltowski, A.J., Castilho, R.F., Grijalba, M.T., Bechara, E.J. & Vercesi A.E. (1996) Effect of inorganic phosphate concentration on the nature of inner mitochondrial membrane alterations mediated by Ca2+ ions. A proposed model for phosphate-stimulated lipid peroxidation. *J Biol Chem. 271*, 2929-2934.

Kowaltowski, A.J. & Vercesi, A.E. (1999) Mitochondrial damage induced by conditions of oxidative stress. *Free Radic. Biol. Med.*, *26*, 463-471.

Kowaltowski, A.J., Castilho, R.F. & Vercesi, A.E. (2001) Mitochondrial permeability transition and oxidative stress. *FEBS Lett.*, *495*, 12-15. Review.

Kranz, R., Lill, R., Goldman, B., Bonnard, G. & Merchant, S. (1998) Molecular mechanisms of cytochrome *c* biogenesis: three distinct systems. *Mol. Microbiol.*, *29*, 383–396.

Kroemer, G., Zamzami, N. & Susin, S.A. (1997) Mitochondrial control of apoptosis. *Immunol. Today*, *18*, 44–51.

Kroemer, G. & Reed, J.C. (2000) Mitochondrial control of cell death. *Nat. Med.*, *6*, 513-519.

Krohn, A.J., Wahlbrink, T. & Prehn, J.H. (1999) Mitochondrial depolarization is not required for neuronal apoptosis. *J. Neurosci.*, *19*, 7394–7404.

Kruman, I. & Mattson, M., (1999) Pivotal role of mitochondrial calcium uptake in neuronal cell apoptosis and necrosis. *J. Neurochem.*, *72*, 529-540.

Larrick, J.W. & Wright, S.C. (1990) Cytotoxic mechanism of tumor necrosis factor-alpha. *FASEB J.*, *4*, 3215-3223. Review

Le Bras, M., Clement, M.V., Pervaiz, S. & Brenner, C. (2005) Reactive oxygen species and the mitochondrial signaling pathway of cell death. *Histol. Histopathol.*, *20*, 205–219.

Lee, H. & Wei, Y. (2000) Mitochondrial role in life and death of the cell. *J. Biomed. Sci.*, *7*, 2–15.

Leist, M., Gantner, F., Jilg, S. & Wendel, A. (1995) Activation of the 55 kDa TNF receptor is necessary and sufficient for TNF-induced liver failure, hepatocyte apoptosis, and nitrite release. *J. Immunol., 154,* 1307-1316.

Leist, M., Single, B., Castoldi, A.F., Kuhnle, S. & Nicotera, P. (1997) Intracellular adenosine triphosphate (ATP) concentration: a switch in the decision between apoptosis and necrosis. *J. Exp. Med., 185,* 1481– 1486.

Lenaz, G., D'Aurelio, M., Merlo Pich, M., Genova, M.L., Ventura, B., Bovina, C., Formiggini, G., Parenti Castelli, G. (2000) Mitochondrial bioenergetics in aging. *Biochim. Biophys. Acta., 1459,* 397-404. Review.

Lennon, S.V., Martin, S.J. & Cotter, T.G. (1991) Dose-dependent induction of apoptosis in human tumour cell lines by widely diverging stimuli. *Cell. Prolif., 24,* 203-214.

Levi, G., Aloisi, F., Ciotti, M.T. & Gallo, V. (1984) Autoradiographic localization and depolarization-induced release of acidic amino acids in differentiating cerebellar granule cell cultures. *Brain Res., 290,* 77–86.

Li, P., Nijhawan, D., Budihardjo, I., Srinivasula, S.M., Ahmad, M., Alnemri, E.S. & Wang, X. (1997): Cytochrome *c* and dATP-dependent formation of Apaf-1/caspase-9 complex initiates an apoptotic protease cascade. *Cell, 91,* 479-489.

Li, L.Y., Luo, X. & Wang, X. (2001) Endonuclease G is an apoptotic DNase when released from mitochondria. *Nature, 412,* 95-99.

Li, W., Srinivasula, S.M., Chai, J., Li, P., Wu, J.W., Zhang, Z., Alnemri, E.S. & Shi, Y. (2002) Structural insights into the pro-apoptotic function of mitochondrial serine protease HtrA2/Omi. *Nat. Struct. Biol., 9,* 436-441.

Lindholm, D., Eriksson, O. & Korhonen, L. (2004) Mitochondrial proteins in neuronal degeneration. *Biochem. Biophys. Res. Commun., 321,* 753-758.

Liu, X., Kim, C.N., Yang, J., Jemmerson, R. & Wang, X. (1996) Induction of apoptotic program in cell-free extracts: requirement for dATP and cytochrome *c*. *Cell, 86,* 147-157.

Logan, D.C. (2006) The mitochondrial compartment. *J. Exp. Botany, 57,* 1225–1243.

Maher, F. & Simpson, I.A. (1994) Modulation of expression of glucose transporters GLUT3 and GLUT1 by potassium and N-methyl-d-aspartate in cultured cerebellar granule neurons. *Mol. Cell. Neurosci., 5,* 369-375.

Majima, E., Shinohara, Y., Yamaguchi, N., Hong, Y.M. & Terada, H. (1994) Importance of loops of mitochondrial ADP/ATP carrier for its transport activity deduced from reactivities of its cysteine residues with the sulfhydryl reagent eosin-5-maleimide. *Biochemistry, 33,* 9530-9536.

Malisan, F. & Testi, R. (2003) Mitochondrial lipids as apoptosis regulators. *Curr. Med. Chem., 10,* 1573-1580.

Malorni, W., Rivabene, R., Santini, M. T., & Donelli, G. (1993). *FEBS Lett.,* 327, 75–78.

Manev, H., Cagnoli, C.M., Atabay, C., Kharlamov, E., Ikonomovic, M.D. & Grayson, D.R. (1995) Neuronal apoptosis in an in vitro model of photochemically induces oxidative stress. *Exp. Neurol., 133,* 198-206.

Markley, H.G., Faillace, L.A. & Mezey, E. (1973) Xanthine oxidase activity in rat brain. *Biochim. Biophys. Acta, 309,* 23-31.

Martin, S.J. & Green, D.R. (1995) Protease activation during apoptosis: death by a thousand cuts? *Cell, 82,* 349-352.

Martin, S.J., Amarante-Mendes, G.P., Shi, L., Chuang, T.H., Casiano, C.A., O'Brien, G.A., Fitzgerald, P., Tan, E.M., Bokoch, G.M., Greenberg, A.H. & Green, D.R. (1996) The cytotoxic cell protease granzyme B initiates apoptosis in a cell-free system by proteolytic processing and activation of the ICE/CED-3 family protease, CPP32, via a novel two-step mechanism. *EMBO J., 15,* 2407–2416.

Martin-Romero, F.J., Garcia-Martin, E. & Gutierrez-Merino, C. (2002) Inhibition of oxidative stress produced by plasma membrane NADH oxidase delays low-potassium-induced apoptosis of cerebellar granule cells. *J. Neurochem., 82,* 705-715.

Martin-Romero, F.J., Gutierrez-Martin, Y., Henao, F. & Gutierrez-Merino, C. (2002) The NADH oxidase activity of the plasma membrane of synaptosomes is a major source of superoxide anion and is inhibited by peroxynitrite. *J. Neurochem., 82,* 604-614.

Marzo, I., Brenner, C., Zamzami, N., Jurgensmeier, J.M., Susin, S.A., Vieira, H.L., Prevost, M.C., Xie, Z., Matsuyama, S., Reed, J.C. & Kroemer, G. (1998) Bax and adenine nucleotide translocator cooperate in the mitochondrial control of apoptosis. *Science, 281,* 2027–2031.

May, J.M., Qu, Z.C. & Whitesell, R.R. (1995) Ascorbate is the major electron donor for a transmembrane oxidoreductase of human erythrocytes. *Biochim. Biophys. Acta, 1238,* 127-136.

McCarthy, N.J., Whyte, M.K., Gilbert, C.S. & Evan, G.I. (1997) Inhibition of Ced-3/ICE-related proteases does not prevent cell death induced by oncogenes, DNA damage, or the Bcl-2 homologue Bak. *J. Cell. Biol. 136,* 215-227.

McCord, R.C. (1985) Neodymium:YAG laser: theoretical and practical considerations. *Int. Ophthalmol. Clin., 25,* 15-19.

Medema, J.P., Scaffidi, C., Kischkel, F.C., Shevchenko, A., Mann, M., Krammer, P.H. & Peter, M.E. (1997) FLICE is activated by association with the CD95 death-inducing signaling complex (DISC). *Embo J.,* 16, 2794–2804.

Minervini, M., Atlante, A., Gagliardi, S., Ciotti, M.T., Marra, E. & Calissano, P. (1997) Glutamate stimulates 2-deoxyglucose uptake in rat cerebellar granule cells. *Brain Res., 768,* 57-62.

Miramar, M.D., Costantini, P., Ravagnan, L., Saraiva, L.M., Haouzi, D., Brothers, G., Penninger, J.M., Peleato, M.L., Kroemer G. & Susin S.A. (2001) NADH oxidase activity of mitochondrial apoptosis-inducing factor. *J. Biol. Chem. 276,* 16391-16398.

Moran, J. & Patel, A.J. (1989) Effect of potassium depolarization on phosphate-activated glutaminase activity in primary cultures of cerebellar granule neurons and astroglial cells during development. *Dev. Brain Res., 46,* 97-105.

Moran, J., Itoh, T., Reddy, U.R., Chen, M., Alnemri, E.S. & Pleasure, D. (1999) Caspase-3 expression by cerebellar granule neurons is regulated by calcium and cyclic AMP. *J. Neurochem., 73,* 568-577.

Nakagawa, Y. (2004a) Role of mitochondrial phospholipid hydroperoxide glutathione peroxidase (PHGPx) as an antiapoptotic factor. *Biol. Pharm. Bull., 27,* 956-960 Review.

Nakagawa, Y. (2004b) Initiation of apoptotic signal by the peroxidation of cardiolipin of mitochondria. *Ann. N. Y. Acad. Sci., 1011,* 177–184.

Nakagawa, T., Shimizu, S., Watanabe, T., Yamaguchi, O., Otsu, K., Yamagata, H., Inohara, H., Kubo, T. & Tsujimoto, Y. (2005) Cyclophilin D-dependent mitochondrial permeability transition regulates some necrotic but not apoptotic cell death. *Nature, 434*, 652–658.

Neame, S.J., Rubin, L.L. & Philpott, K.L. (1998) Blocking cytochrome c activity within intact neurons inhibits apoptosis. *J. Cell. Biol., 142*, 1583–1593.

Neupert, W. (1997) Protein import into mitochondria. *Annu. Rev. Biochem., 66*, 863–917.

Nicholls, D.G. & Ferguson, S.J. (2002) Bioenergetics 2, Academic Press, London.

Nicholson, D.W. (1999) Caspase structure, proteolytic substrates, and function during apoptotic cell death. *Cell Death Diff., 6*, 1028-1042.

Nicotera, P., Leist, M. & Ferrando-May, E. (1998) Intracellular ATP, a switch in the decision between apoptosis and necrosis. *Toxicol. Lett., 102-103*, 139-142. Review.

Nicotera, P., Leist, M., Fava, E., Berliocchi, L. & Volbracht, C. (2000) Energy requirement for caspase activation and neuronal cell death. *Brain Pathol., 10*, 276–282.

Nicotera, P. & Melino, G. (2004) Regulation of the apoptosis-necrosis switch. *Oncogene 23*, 2757-2765.

Opie, L.H. (1992) Cardiac metabolism-emergence, decline, and resurgence: Part I. *Cardiovasc. Res., 26*, 721–733.

Oppenheim, R.W. (1991) Cell death during development of the nervous system. *Annu. Rev. Neurosci., 14*, 453-501 Review.

Orrenius, S. (2004) Mitochondrial regulation of apoptotic cell death. *Toxicol. Lett., 149*, 19-23 Review.

Ott, M., Robertson, J.D., Gogvadze, V., Zhivotovsky, B. & Orrenius, S. (2002) Cytochrome c release from mitochondria proceeds by a two-step process. *Proc. Natl Acad. Sci. U.S.A., 99*, 1259–1263.

Palade, G. (1952) The fine structure of mitochondria. *Anatomical Record, 114*, 427–451.

Palomba, L., Sestili, P., Cattabeni, F., Azzi, A. & Cantoni, O. (1996) Prevention of necrosis and activation of apoptosis in oxidatively injured human myeloid leukemia U937 cells. *FEBS Lett., 390*, 91-94.

Pan, Z., Voehringer, D.W. & Meyn, R.E. (1999) analysis of redox regulation of cytochrome *c*-induced apoptosis in a cell-free system. *Cell Death Differ., 6*, 683–688.

Parrish, J., Li, L., Klotz, K., Ledwich, D., Wang, X. & Xue, D. (2001) Mitochondrial endonuclease G is important for apoptosis in *C. elegans*. *Nature, 412*, 90-94.

Passarella, S., Atlante, A., Valenti, D. & de Bari, L. (2003) The role of mitochondrial transport in energy metabolism. *Mitochondrion, 2*, 319–343.

Patterson, S.D., Spahr, C.S., Daugas, E., Susin, S.A., Irinopoulou, T., Koehler, C. & Kroemer, G. (2000) Mass spectrometric identification of proteins released from mitochondria undergoing permeabilità transition. *Cell. Death Differ., 7*, 137–144.

Pereverzev, M.O., Vygodina, T.V., Konstantinov, A.A. & Skulachev, V.P. (2003) Cytochrome *c*, an ideal antioxidant. *Biochem. Soc. Trans., 31*, 1312–1315.

Perkins, G.A., Renken, C.W., Frey, T.G. & Ellisman, M.H. (2001) Membrane architecture of mitochondria in neurons of the central nervous system. *J. Neurosci. Res. 66*, 857–865.

Peters, J.M. (1994) Proteasome:protein degradation machines of the cell. *Trends Biochem. Sci., 19*, 377-382.

Petrosillo, G., Ruggiero, F.M. & Paradies, G. (2003) Role of reactive oxygen species and cardiolipin in the release of cytochrome *c* from mitochondria. *FASEB J., 17*, 2202-2208.

Plaetzer, K., Kiesslich, T., Krammer, B. & Hammerl, P. (2002) Characterization of the cell death modes and the associated changes in cellular energy supply in response to AlPcS4-PDT. *Photochem. Photobiol. Sci., 1,* 172-177.

Polster, B.M. & Fiskum, G. (2004) Mitochondrial mechanisms of neural cell apoptosis. *J. Neurochem., 90,* 1281-1289.

Priault, M., Chaudhuri, B., Clow, A., Camougrand, N. & Manon, S. (1999) Investigation of bax-induced release of cytochrome c from yeast mitochondria permeability of mitochondrial membranes, role of VDAC and ATP requirement. *Eur. J. Biochem., 260,* 684–691.

Qin, Z.H., Wang, Y., Kikly, K.K., Sapp, E., Kegel, K.B., Aronin, N. & DiFiglia, M. (2001) Pro-caspase-8 is predominantly localized in mitochondria and released into cytoplasm upon apoptotic stimulation. *J. Biol. Chem., 276,* 8079-8086.

Rabizadeh, S., Gralla, E.B., Borchelt, D.R., Gwinn, R., Selverstone, V.J., Sisodia, S., Wong, P., Lee, M., Hahn, H. & Bredesen, D.E. (1995) Mutations associated with amyotrophic lateral sclerosis convert superoxide dismutase from an antiapoptotic gene to a proapoptotic gene: studies in yeast and neural cells. *Proc. Natl. Acad. Sci. U.S.A., 92,* 3024-3028.

Reed, J.C. (1997) Double identity for proteins of the Bcl-2 family. *Nature, 387,* 773–776.

Reed, J.C. (2000) Mechanisms of apoptosis. *Am. J. Pathol., 157,* 1415-1430.

Renehan, A.G., Booth, C. & Potten, C.S. (2001) What is apoptosis, and why is it important? *BMJ, 322*, 536-1538.

Resink, A., Hack, N., Boer, G.J. & Balàsz, R. (1994) Growth conditions differentially modulate the vulnerability of developing cerebellar granule cells to excitatory amino acids. *Brain Res., 655*, 222-232.

Sadoul, R., Fernandez, P.A., Quiquerez, A.L., Martinou, I., Maki, M., Schroter, M., Becherer, J.D., Irmler, M., Tschopp, J. & Martinou, J.C. (1996) Involvement of the proteasome in the programmed cell death of NGF-deprived sympathetic neurons. *EMBO J., 15,* 3845-3852.

Sadoul, R. (1998) Bcl-2 family members in the development and degenerative pathologies of the nervous system. Cell Death Differ., 5, 805-815 Review.

Saleh, A., Srinivasula, S.M., Acharya, S., Fishel, R. & Alnemri, E.S. (1999) Cytochrome *c* and dATP-mediated oligomerization of Apaf-1 is a prerequisite for procaspase-9 activation. *J. Biol. Chem*., 274, 17941–17945.

Samali, A., Zhivotovsky, B., Jones, D.P. & Orrenius, S. (1998) Detection of pro-caspase-3 in cytosol and mitochondria of various tissues. *FEBS Lett., 431,* 167-169.

Sarin, A., Williams, M.S., Alexander-Miller, M.A., Berzofsky, J.A., Zacharchuk, C.M. & Henkart, P.A. (1997) Target cell lysis by CTL granule exocytosis is independent of ICE/Ced-3 family proteases. *Immunity, 6*, 209-215.

Scaffidi, C., Krammer, P.H. & Peter, M.E. (1999) Isolation and analysis of components of CD95 (APO-1/Fas) death-inducing signaling complex. *Methods, 17,* 287–291.

Schulz, J.B., Weller, M. & Klockgether, T. (1996) Potassium deprivation induced apoptosis of cerebellar granule neurons: a sequential requirement for new mRNAand protein synthesis, ICE-like protease activity, and reactive oxygen species. *J. Neurosci., 16,* 4696–4706.

Scorrano, L., Ashiya, M., Buttle, K., Weiler, S., Oakes, S.A., Mannella, C.A. & Korsmeyer, S.J. (2002) A distinct pathway remodels mitochondrial cristae and mobilizes cytochrome c during apoptosis. *Dev. Cell., 2,* 55–67.

Sharpe, J.C., Arnoult, D. & Youle, R.J. (2004) Control of mitochondrial permeability by Bcl-2 family members. *Biochim. Biophys. Acta, 1644,* 107–113.

Shimizu, S., Narita, M. & Tsujimoto, Y. (1999) Bcl-2 family proteins regulate the release of apoptogenic cytochrome c by the mitochondrial channel VDAC. *Nature, 399,* 483–487.

Shohat, G., Spivak-Kroizman, T., Eisenstein, M. & Kimchi, A. (2002) The regulation of death-associated protein (DAP) kinase in apoptosis. *Eur. Cytokine Netw., 13,* 398–400.

Skulachev, V. (1998) Cytochrome *c* in the apoptotic and antioxidant cascades. *FEBS Lett., 423,* 275–280.

Slee, E.A., Harte, M.T., Kluck, R.M., Wolf, B.B., Casiano, C.A., Newmeyer, D.D., Wang, H.G., Reed, J.C., Nicholson, D.W., Alnemri, E.S., Green, D.R. and Martin, S.J. (1999) Ordering the cytochrome c-initiated caspase cascade: hierarchical activation of caspases-2, -3, -6, -7, -8, and -10 in a caspase-9- dependent manner. *J. Cell Biol., 144,* 281–292.

Slee, E.A., Keogh, S.A. & Martin, S.J. (2000) Cleavage of BID during cytotoxic drug and UV radiation-induced apoptosis occurs downstream of the point of Bcl-2 action and is catalyzed by Caspase-3: a potential feedback loop for amplification of apoptosis-associated mitochondrial cytochrome *c* release. *Cell Death Diff. 7,* 556–565.

Slee, E.A., Adrain, C. & Martin, S.J. (2001) Executioner caspases-3, -6, and -7 perform distinct, non-redundant, roles during the demolition phase of apoptosis. *J. Biol. Chem., 276,* 7320–7326.

Sperandio, S., de Belle, I. & Bredesen, D.E. (2000) An alternative, nonapoptotic form of programmed cell death. *Proc. Nat. Acad. Sci. U.S.A., 97,* 14376-14381.

Srinivasula, S.M., Hegde, R., Saleh, A., Datta, P., Shiozaki, E., Chai, J., Lee, R.A., Robbins, P.D., Fernandes-Alnemri, T., Shi, Y. & Alnemri, E.S. (2001) A conserved XIAP-interaction motif in caspase-9 and Smac/DIABLO regulates caspase activity and apoptosis. *Nature, 410,* 112-116.

Stennicke, H.R., Jurgensmeier, J.M., Shin, H., Deveraux, Q., Wolf, B.B., Yang, X., Zhou, Q., Ellerby, H.M., Ellerby, L.M., Bredesen, D., Green, D.R., Reed, J.C., Froelich, C.J. & Salvesen, G.S. (1998) Pro-caspase-3 is a major physiologic target of caspase-8. *J. Biol. Chem., 273,* 27084–27090.

Sugioka, K., Nakano, M., Totsune Nakano, H., Minatami, H., Tero Kubota, S. & and Ikegami, Y. (1988) Mechanism of $O_2^-$ generation in reduction and oxidation cycle of ubiquinones in a model of mitochondrial electron transport systems. *Biochim. Biophys. Acta, 936,* 377-385.

Susin, S.A., Lorenzo, H.K., Zamzami, N., Marzo, I., Snow, B.E., Brothers, G.M., Mangion, J., Jacotot, E., Costantini, P., Loeffler, M., Larochette, N., Goodlett, D.R., Aebersold, R., Siderovski, D.P., Penninger, J.M. & Kroemer, G. (1999a) Molecular characterization of mitochondrial apoptosis-inducing factor. *Nature, 397,* 441-446.

Susin, S.A., Lorenzo, H.K., Zamzami, N., Marzo, I., Brenner, C., Larochette, N., Prevost, M.C., Alzari, P.M. & Kroemer, G. (1999b) Mitochondrial release of caspase-2 and -9 during the apoptotic process. *J. Exp. Med., 189,* 381–394.

Susin, S.A., Daugas, E., Ravagnan, L., Samejima, K., Zamzami, N., Loeffler, M., Costantini, P., Ferri, K.F., Irinopoulou, T., Prevost, M.C., Brothers, G., Mak, T.W., Penninger, J.,

Earnshaw, W.C. & Kroemer, G. (2000) Two distinct pathways leading to nuclear apoptosis. *J. Exp. Med., 192*, 571-580.

Suzuki, Y., Imai, Y., Nakayama, H., Takahashi, K., Takio, K. & Takahashi, R. (2001) A serine protease, HtrA2, is released from the mitochondria and interacts with XIAP, inducing cell death. *Mol. Cell., 8,* 613-621.

Tan, S., Sagara, Y., Liu, Y., Maher, P. & Schubert, D. (1998) The regulation of reactive oxygen species production during programmed cell death. *J. Cell Biol., 141,* 1423–1432.

Tatton, W.G., Wadia, J.S., Ju, W.Y., Chalmers-Redman, R.M. & Tatton, N.A. (1996) (-)-Deprenyl reduces neuronal apoptosis and facilitates neuronal outgrowth by altering protein synthesis without inhibiting monoamine oxidase. *J. Neural Transm., 48,* 45– 59.

Tatton, W.G. & Chalmers-Redman, R.M. (1998) Mitochondria in neurodegenerative apoptosis: an opportunity for therapy? *Ann. Neurol., 44,* S134–S141.

Temple, S. (2001) The development of neural stem cells. *Nature, 414*, 112-117 Review.

Thangnipon, W., Kingsbury, A., Webb, M. & Balazs, R. (1983) Observations on rat cerebellar cells in vitro: influence of substratum, potassium concentration and relationship between neurones and astrocytes. *Brain Res., 313,* 177-189.

Tobias, D.J., Mar, W., Blasie, J.K. & Klein, M.L. (1996) Molecular dynamics simulations of a protein on hydrophobic and hydrophilic surfaces. *Biophys. J., 71,* 2933-2941.

Tsujimoto, Y. & Shimizu, S. (2000) Bcl-2 family: life-or-death switch. *FEBS Lett., 466,* 6-10 Review.

Tymianski, M., Wang, L.Y. & MacDonald, J.F. (1994) Alteration of neuronal calcium homeostasis and excitotoxic vulnerability by chronic depolarization. *Brain Res., 648***,** 291-295.

Valencia, A. & Moran, J. (2001) Role of oxidative stress in the apoptotic cell death of cultured cerebellar granule neurons. *J. Neurosi. Res., 64,* 284–297.

Valencia, A. & Morán, J. (2004) Reactive oxygen species induce different cell death mechanisms in cultured neurons. *Free Rad. Biol. Med., 36,* 1112-1125

Valle, V.G., Fagian, M.M., Parentoni, L.S., Meinicke, A.R. & Vercesi, A.E. (1993) The participation of reactive oxygen species and protein thiols in the mechanism of mitochondrial inner membrane permeabilization by calcium plus prooxidants. *Arch. Biochem. Biophys., 307*, 1-7.

Van de Craen, M., Declercq, W., van den Brande, I., Fiers, W. & Vandenabeele, P. (1999) The proteolytic procaspase activation network: an in vitro analysis. *Cell Death Diff., 6,* 1117-1124.

van Loo, G., Schotte, P., van Gurp, M., Demol, H., Hoorelbeke, B., Gevaert, K., Rodriguez, I., Ruiz-Carrillo, A., Vandekerckhove, J., Declercq, W., Beyaert, R. & Vandenabeele, P. (2001) Endonuclease G: a mitochondrial protein released in apoptosis and involved in caspaseindependent DNA degradation. *Cell. Death Differ., 8,* 1136–1142.

Vercammen, D., Beyaert, R., Denecker, G., Goossens, V., Van Loo, G., Declercq, W., Grooten, J., Fiers, W. & Vandenabeele, P. (1998a) Inhibition of caspases increases the sensitivity of L929 cells to necrosis mediated by tumor necrosis factor. *J. Exp. Med., 187,* 1477-1485.

Vercammen, D., Brouckaert, G., Denecker, G., Van de Craen, M., Declercq, W., Fiers, W. & Vandenabeele, P. (1998b) Dual signaling of the Fas receptor: initiation of both apoptotic and necrotic cell death pathways. *J. Exp. Med., 188,* 919-930

Vercesi, A.E., Kowaltowski, A.J., Grijalba, M.T., Meinicke, A.R. & Castilho, R.F. (1997) The role of reactive oxygen species in mitochondrial permeability transition. *Biosci. Rep., 17,* 43-52 Review.

Verhagen, A.M., Ekert, P.G., Pakusch, M., Silke, J., Connolly, L.M., Reid, G.E., Moritz, R.L., Simpson, R.J. & Vaux, D.L. (2000) Identification of DIABLO, a mammalian protein that promotes apoptosis by binding to and antagonizing IAP proteins. *Cell, 102,* 43-53.

Verhagen, A.M., Silke, J., Ekert, P.G., Pakusch, M., Kaufmann, H., Connolly, L.M., Day, C.L., Tikoo, A., Burke, R., Wrobel, C., Moritz, R.L., Simpson, R.J. & Vaux, D.L. (2002) HtrA2 promotes cell death through its serine protease activity and its ability to antagonize inhibitor of apoptosis proteins. *J. Biol. Chem., 277,* 445-454.

Verrier, F., Deniaud, A., Lebras, M., Metivier, D., Kroemer, G., Mignotte, B., Jan, G. & Brenner, C. (2004) Dynamic evolution of the adenine nucleotide translocase interactome during chemotherapyinduced apoptosis. *Oncogene, 23,* 8049–8064.

Vieira, H. L., Haouzi, D., El Hamel, C., Jacotot, E., Belzacq, A.S., Brenner, C. & Kroemer, G. (2000) Permeabilization of the mitochondrial inner membrane during apoptosis: impact of the adenine nucleotide translocator. *Cell. Death Differ. 7,* 1146–1154.

Wang, X. (2001) The expanding role of mitochondria in apoptosis. *Genes Dev., 15,* 2922-2933.

Waterhouse, N.J., Goldstein, J.C., von Ahsen, O., Schuler, M., Newmeyer, D.D. & Green, D.R. (2001) Cytochrome *c* maintains mitochondrial transmembrane potential and ATP generation after outer mitochondrial membrane permeabilization during the apoptotic process. *J. Cell. Biol., 153,* 319–328.

Waterhouse, N.J., Ricci, J.-E. & Green, D.R. (2002) And all of a sudden it's over: mitochondrial outer-membrane permeabilization in apoptosis. *Biochimie, 84,* 113-121.

Wells, W.W. & Xu, D.P. (1994) Dehydroascorbate reduction. *J. Bioenerg. Biomembr.*, 26, 369-377

Wigdal, S.S., Kirkland, R.A., Franklin, J.L. & Haak-Frendscho, M. (2002) Cytochrome c release precedes mitochondrial membrane potential loss in cerebellar granule neuron apoptosis: lack of mitochondrial swelling. *J. Neurochem., 82*, 1029–1038.

Wilkinson, K. (1997) Regulation of ubiquitin-dependent process by deubiquitinating enzymes. *FASEB J., 11,* 1245-1246.

Wolf, B.B., Schuler, M., Li, W., Eggers-Sedlet, B., Lee, W. Tailor, P., Fitzgerald, P., Mills, G.B. & Green, D.R. (2001) Defective cytochrome *c*-dependent caspase activation in ovarian cancer cell lines due to diminished or absent apoptotic protease activating factor-1 activity. *J. Biol. Chem., 276,* 34244–34251;

Wong, G.H. (1995) Protective roles of cytokines against radiation: induction of mitochondrial MnSOD. *Biochim. Biophys. Acta., 1271,* 205-209.

Xia, T., Jiang, C., Li, L., Wu, C., Chen, Q. & Liu, S.S. (2002) A study on permeability transition pore opening and cytochrome c release from mitochondria, induced by caspase-3 in vitro. *FEBS Lett., 510,* 62–66.

Yang, J., Liu, X., Bhalla, K., Kim, C.N., Ibrado, A.M., Cai, J., Peng, T.I., Jones, D.P. and Wang, X. (1997) Prevention of apoptosis by Bcl-2: release of cytochrome c from mitochondria blocked. *Science, 275,* 1129-1132.

Yu, C., Subler, M., Rahmani, M., Reese, E., Krystal, G., Conrad, D., Dent, P. & Grant, S. (2003) Induction of apoptosis in BCR/ABL+ cells by histone deacetylase inhibitors involves reciprocal effects on the RAF/MEK/ERK and JNK pathways. *Cancer Biol. Ther., 2,* 544-551.

Yu, C., Rahmani, M., Almenara, J., Sausville, E.A., Dent, P. & Grant, S. (2004) Induction of apoptosis in human leukemia cells by the tyrosine kinase inhibitor adaphostin proceeds through a RAF-1/MEK/ERK- and AKT-dependent process. *Oncogene, 23,* 1364-1376.

Zamzami, N. & Kroemer, G. (2001) The mitochondrion in apoptosis: how Pandora's box opens. *Nat. Rev. Mol. Cell. Biol., 2,* 67–71.

Zhivotovsky, B., Orrenius, S., Brustugun, O.T. & Doskeland, S.O. (1998) Injected cytochrome c induces apoptosis. *Nature, 391,* 449-450.

Zhuang, S., Demirs, J.T. & Kochevar, I.E. (2000) p38 mitogen-activated protein kinase mediates bid cleavage, mitochondrial dysfunction induced by singlet oxygen but not by hydrogen peroxide. *J. Biol. Chem., 275,* 25939-25948.

Zhuang, S., Demris, J.T. & Kochevar, I.E. (2001) Protein kinase C inhibits singlet oxygen-induced apoptosis by decreasing caspase-8 activation. *Oncogene, 20,* 6764-6776.

Zou, H., Li, Y., Liu, X. & Wang, X. (1999) An APAF-1 cytochrome c multimeric complex is a functional apoptosome that activates procaspase-9. *J. Biol. Chem., 274,* 11549–11556.

In: Neuronal Cell Apoptosis Research
Editor: Relif J. Shandling, pp. 47-74
ISBN 1-60021-456-8


Chapter 2

# Fibrillar Aggregates of Amyloidogenic Peptides: New Insight into the Neuroprotective Effects of Neuropeptides on Fibril-Induced Neuronal Cell Death

***Satomi Onoue*[1,2,*] *and Takehiko Yajima*[1]**
[1]Department of Analytical Chemistry, Faculty of Pharmaceutical Sciences, Toho University, Funabashi, Chiba 274-8510, Japan;
[2]Analytical Research and Development, Pfizer Global Research and Development, Nagoya Laboratories, Pfizer Japan Inc., Taketoyo, Aichi 470-2393, Japan.

## Abstract

The generation of insoluble peptide/protein fibrils is well-confirmed in amyloidosis, complex disorders characterized by the polymerization and aggregation of normally innocuous and soluble proteins or peptides followed by extracellular insoluble fibrils with resistance to peptidases. In this investigation, we demonstrated that pathologically-related amyloidogenic proteins/peptides, including prion protein fragment (PrP), ß-amyloid(1–42), and amylin, generated the insoluble fibrils, and it induced delayed cell death in neuronal PC-12 cells. In addition to these pathogenic peptides/proteins, some therapeutic peptides, including glucagon and salmon calcitonin, also had an ability to form gel-like fibrillar aggregates as observed by transmission electron microscopy. On the basis of circular dichroic analysis, the peptides-derived fibrils were found to share a distinct conformational feature in the richness of the ß-sheet structure, despite having no sequential homology among these peptides. All tested peptide fibrils, including the

* Correspondence concerning this article should be addressed to Satomi Onoue, Analytical Research and Development, Pfizer Global Research and Development, Nagoya Laboratories, Pfizer Japan Inc., 5-2 Taketoyo, Aichi 470-2393, Japan. Tel: +81-569-74-4855; Fax: +81-569-74-4748, E-mail. onoue@fukai.or.jp.

pathologically-related and therapeutic amyloidogenic proteins/peptides, caused neurotoxic effects with the significant activation of caspase-3, an apoptotic enzyme, in PC12.

Pituitary adenylate cyclase-activating polypeptide (PACAP) and vasoactive intestinal peptide (VIP), peptide transmitters in the central and peripheral nervous systems, potently stimulate neuritogenesis and survival of neuronal cells in the central nervous system. Concomitant application of these neuropeptides with peptides aggregates, such as aged PrP(106–126) and glucagon, inhibited the delayed death of neuron-like PC12 cells. In particular, PACAP27 inhibited the neurotoxicity of peptides fibrils at low concentrations (>$10^{-15}$ M), characterized by the deactivation of fibrils-stimulated caspase-3. The neuroprotective effect of PACAP27 was antagonized by the selective PKA inhibitor, H89, or MAP kinase inhibitor, U0126, suggesting that PACAP27 attenuates fibrils-induced delayed neurotoxicity in PC12 cells by activating both PKA and MAP kinases mediated by PAC1 receptor.

**Keywords:** PACAP; VIP; PC12 cells; PrP(106–126); Glucagon; Apoptosis; Caspase.

## INTRODUCTION

Conformational diseases, including Alzheimer's, Parkinson's and other late-onset neurodegenerative diseases, arise from the conformationally driven aggregation of individual proteins [1]. Especially, the generation of insoluble peptide/protein fibrils is well-confirmed in amyloidosis, and the extracellular insoluble fibrils were found to be resistant to peptidases [2,3]. There are a lot of proteins forming amyloid fibrils (Table 1) [4-6], which include ß-amyloid (Aß) in Alzheimer's disease [7], amylin in type II (non-insulin-dependent) diabete mellitus [8], prion protein (PrP) in Creutzfeldt-Jakob disease and spongiform encephalopathy [9], and polyglutamine in Huntington's disease [10]. In addition to these amyloidosis-related peptides/protein, some therapeutic peptides/proteins, such as glucagon, insulin, and calcitonin, have ability to generate fibril aggregates under specific conditions (Table 1) [11]. These insoluble products are attributed to the formation of partially unfolded intermediates with an exposed hydrophobic region that drives the aggregation towards the pharmaceutically undesirable form [12]. These amyloidogenic protein/peptide fibrils share a distinct conformational feature in the richness of the ß-sheet structure [13]. In addition, there are similar characteristics of polarity, hydrophobicity, and the size of side-chain among certain segments containing 10–15 amino acid residues of amyloid-forming peptides such as insulin, Aß, and amylin, and these factors were indicative of a consensus sequence as a recognition motif of Congo red, a specific dye for amyloidogenic protein/peptide fibrils with ß-sheet dependency [14,15].

Prion diseases are characterized by the accumulation of large aggregates of the PrP in both human and animal brains, and they are unusual fetal neurodegenerative disorders including Creutzfeldt-Jakob disease in humans, scrapie in sheep and goats, and spongiform encephalopathy in cattle [9]. The prion protein fragment, PrP(106–126), corresponding to a putative transmembrane region of cellular PrP, is believed to be one of the key regions where conformational changes are initiated leading to the conversion of $PrP^{c}$ to $PrP^{sc}$ [16]. This peptide shows a remarkable conformational polymorphism, acquiring different secondary structures in various environments, and it has been shown to induce cell death in neurons likely via the programmed cell death pathway. In this investigation, we have characterized and compared the physicochemical and physiological properties of PrP(106–126) and other amyloidogenic peptides/proteins, such as Aß(1–42), amylin, glucagon, and salmon calcitonin, with the use of analytical and biophysical techniques including circular dichroism (CD), transmission electron microscopy (TEM), ß-sheet-imaging probes, and computational method. Here, we have demonstrated that the fibrils of peptides/proteins possess the same conformational properties and cytotoxic apoptotic signaling pathways by activating caspases.

**Table 1. Amyloidogenic peptides/proteins.**

| *Amyloidosis-related peptides/proteins* | *Disease* |
|---|---|
| α-Synuclein | Parkinson's disease |
| Amylin | Type II diabetes |
| Androgen receptor | Spinobulbar muscular atrophy |
| Apolipoprotein AI | Familial amyloid polyneuropathy III |
| Ataxins | Spinocerebellar ataxias |
| Atrial natriuretic factor | Atrial amyloidosis |
| Atrophin-1 | Dentatorubral pallidoluysian atrophy |
| $\beta_2$-Microglobulin | Haemodialysis-related amyloidosis |
| β-Amyloid | Alzheimer's disease |
| BRI protein | Familial dementia |
| Calcitonin | Medullary carcinoma-associated amyloidosis |
| CFTR protein | Cystic fibrosis |
| Cystatin C | Hereditary cerebral amyloid angiopathy |
| Expanded Glu repeats of huntingtin | Huntington's disease |
| Fibrinogen α-A chain | Hereditary renal amyloidosis |
| Gelsolin | Finnish hereditary systemic amyloidosis |
| Heat shock proteins | Progressive supranuclear palsy |
| Immunoglobulin heavy chain | AH amyloidosis |
| Immunoglobulin light chain | AL amyloidosis |
| Keratin | Primary localized cutaneous amyloidosis |
| Lactoferrin | Familial cornea amyloidosis |
| Lysozyme | Hereditary non-neuropathic systemic amyloidosis |
| Prion | Spongiform encephalopathies |
| Prolactin | Ageing pituitary, Prolactinomas |
| Serum amyloid A | Secondary systemic amyloidosis |
| Superoxide dismutase 1 | Amyotrophic lateral sclerosis |
| TATA box-binding protein | Spinocerebellar ataxias |
| Tau | Fronto-temporal dementias |
| Transthyretin | Senile systemic amyloidosis |
| *Amyloidogenic therapeutic peptides/proteins* | *Clinical application* |
| Glucagon | Hypoglycemia |
| Glucagon-like peptide-1 derivative | Diabetes |
| rh Growth hormone | Growth hormone deficiency |
| Insulin | Diabetes |
| Salmon caclitonin | Osteoporosis |

We also provide further insights into the neuroprotective effects of pituitary adenylate cyclase-activating polypeptide (PACAP) [17-19] and vasoactive intestinal peptide (VIP) [20] in the neuronal cells. PACAP and VIP are the major peptide transmitters in the central and peripheral nervous systems, performing a broad spectrum of biological functions affecting both natural and acquired immunity, metabolic processes, exocrine and endocrine secretions, cell differentiation, and smooth muscle relaxation [21,22]. There are two forms of mammal PACAP, PACAP38 and a shorter peptide with the same N-terminal 27 residue, PACAP27, and they have been shown to have the same biological function and receptor-binding activity [23]. Recently, PACAP and VIP have been shown to act as survival factors against ischemia and glutamate-induced neurotoxicities [24,25], and *in vitro* studies have demonstrated that they can stimulate neurite outgrowth [26]. These observations prompted us to investigate the effect of neuropeptides, including PACAP and VIP, on the fibrils-induced neurotoxicity using neuron-like PC12 cells, in which the expression of PACAP-specific PAC1 receptor was confirmed [27]. In addition, we evaluated how PACAP/VIP influence the cytotoxicity level, characterized by caspase cascades and several protein kinase signaling pathways.

## EXPERIMENTAL

### 1. Chemicals

The neuropeptides and amyloidogenic peptides/proteins were chemically synthesized by the solid-phase strategy employing optimal side-chain protection as reported previously [28]. Peptides were removed from the resin by HF treatment, and the cleaved products were purified using C-18 column chromatography (Fuji Silysia, Aichi, Japan). The purity (>98%) was checked by reverse-phase HPLC using a TSK-gel ODS-120T column (Tosoh, Tokyo, Japan), and the homogeneity was assessed by amino acid analyses using an amino acid analyzer L-8500 (Hitachi, Tokyo, Japan). Their molecular weights were confirmed on a matrix-assisted laser desorption ionization-time of flight mass spectrometer (Kratos, Manchester, UK).

WST-8 (4-[26]-1, 3-benzene disulfonate sodium salt) was purchased from Dojindo (Kumamoto, Japan). Dibutyryl-cAMP (db-cAMP), 12-*O*-tetradecanoyl-phorbol-13-acetate (TPA), U0126 (Bis[amino[(2-aminophenyl)thio]methylene]butanedinitrile) and H-89 (N-(2-[*p*-bromocinnamylamino]ethyl)-5-isoquinolinesulfonamide) were purchased from Sigma (St. Louis, MO). Myr-ψPKC (myristoyl-Gly-Arg-Arg-Asn-Ala-Ile-His-Asp-Ile), Ac-DEVD-CHO and Z-VAD-FMK were purchased from ProMega (Madison, WI, USA).

### 2. Prediction of Secondary Structure

The α-helix, ß-sheet, and turn propensities were calculated by the Chou-Fasman secondary prediction algorithm [29,30] using the program DNASIS-Mac.

## 3. Aging Treatment of Peptides

Glucagon, salmon calcitonin, amylin, Aß(1–42), and PrP(106–126) were dissolved in 0.01 M HCl at the concentration indicated in the text, and incubated at 37°C. Then, the solution was diluted to the required concentration with serum-free DMEM containing 2 μM insulin.

## 4. Circular Dichroism Analysis of Amyloidogenic Peptides

Aged preparations of peptides were dissolved in 20 mM Tris-HCl buffer (pH 7.4) or 50% methanol (MeOH)/20 mM Tris-HCl buffer, and circular dichroism (CD) spectra (average of ten scans) were collected from samples (2 ml) at 0.5 nm intervals between wavelengths of 200 and 400 nm using a Jasco model J-720 spectropolarimeter (Tokyo, Japan). Samples were incubated at room temperature and a baseline spectrum was subtracted from the collected data.

## 5. Congo Red Binding Assay

Congo red-reactive fibrils were measured as described previously [31]. Aged preparations of peptides were adjusted to a concentration of 1 mg/ml, and 40 μl of each dilution was added to 960 μl of 25 μM Congo Red in 20 mM potassium phosphate buffer (NaPB, pH 7.4) containing 150 mM NaCl. After a 30-min incubation, absorbance was read at 540 and 477 nm, and the concentration of bound Congo red ($c_b$) was calculated from the equation, $c_b$=(A540/25,295 – A477/46,306).

## 6. Transmission Electron Microscopy (TEM)

An aliquot (2 μl) of the peptide gels or solution was placed on a carbon-coated Formvar 200 mesh nickel grid. The sample was allowed to stand for 15–30 s, and then any excess solution was removed by blotting. The samples were negatively stained with 2% (w/v) uranyl acetate and allowed to dry. The samples were then visualized under a Hitachi H-7000 transmission electron microscope operating at 75 kV.

## 7. Size Exclusion Chromatography (SEC)

Aged preparations of peptides were fractionated on a Zorbax GF-250 column (Agilent Technologies, Palo Alto, CA, USA) at 25°C using a Shimadzu LC-10A HPLC system. The column was equilibrated with the mobile phase (20 mM citrate buffer containing 130 mM NaCl, pH 3.0), and peptides were eluted under constant flow at 1.0 ml/min and monitored at 220 nm. The column was calibrated with blue dextran (2,000,000 Da) and a series of molecular weight protein standards including, sweet potato ß-amylase (200,000 Da), human serum albumin (66,500 Da), chicken albumin (45,000 Da), human growth hormone (22,125 Da), porcine insulin (5,777 Da), and buserelin (1239 Da).

## 8. Cell Cultures

Rat pheochromocytoma (PC12) cells were obtained from the RIKEN Cell Bank (Ibaraki, Japan). The PC12 cells were cultured in Dulbecco's modified Eagle's medium (DMEM; Sigma) supplemented with 5% (v/v) horse serum (HS; Gibco-BRL, Grand Island, NY) and 5% (v/v) newborn calf serum (CS; Gibco-BRL) as described previously [27]. The cultures were maintained in 5% $CO_2$/95% humidified air at 37°C.

## 9. Measurement of Lactate Dehydrogenase (LDH) Activity

PC12 cells were seeded at $10^4$ cells/well in 96-well plates coated with type I collagen (Becton Dickinson Labware) 1 day before the experiment and cultured in serum-free DMEM supplemented with 2 μM insulin. Peptide fibril was added to the cultures at the various concentrations with or without neuropeptides or other reagents. The extent of cell death was assessed by measurement of the activity of LDH released from the dead cells, as the loss of cell membrane integrity was observed in both necrotic and apoptotic cells. LDH activity in the culture medium was determined using a commercially available kit (Wako, Osaka, Japan) according to the manufacturer's protocol.

## 10. Cell Viability Assessment Using WST-8

In addition to LDH measurements in the medium, cell mortality was also assayed by WST-8 conversion [32]. A value of 10 μl of WST-8 (5 mM WST-8, 0.2 mM 1-methoxy-5-methylphenazinium methylsulfate, and 150 mM NaCl) was added to each well and the reaction was continued for 4 h at 37°C. The absorbance of the sample at 450 nm was measured using a microplate reader (BIO-TEK, Winooski, VT) with a reference wavelength of 720 nm.

## 11. Measurement of Caspase-3-Like Activity

The caspase-3-like activity in culture was measured with the use of an Apo-ONE™ Homogeneous Caspase-3/7 Assay Kit (Promega) according to the manufacturer's instructions.

Briefly, the cells ($5 \times 10^4$ cells/well) in type I collagen-coated 96-well plates (Becton Dickinson Labware) were rinsed twice with phosphate-buffered saline (PBS). The cultures were incubated with or without the indicated stimulants in DMEM (50 μl) at 37°C in an atmosphere of 95% air and 5% $CO_2$. The cells were lysed in 50 μl of Homogeneous Caspase-3/7 Buffer containing the caspase-3 substrate Z-DEVD-Rhodamine 110, and the cell lysates were incubated for 12 h at room temperature. After incubation, the cell lysates (50 μl) were diluted with 2 ml of PBS, and the fluorescence (excitation at 480 nm and emission at 535 nm) was measured with an RF-5000 spectrofluorophotometer (Shimadzu, Tokyo, Japan).

## 12. Reverse Transcriptional PCR (RT-PCR) Amplification

The RT-PCR experiments were performed as described in our previous report [27], and the primers for the PACAP/VIP receptors (PAC1, VPAC1, and VPAC2) were as follows: 5'- and 3'-primers for PAC1 (GenBank accession no. Z23279 for the basic receptor, Z23273 for hip, Z23274 for hop1, Z23275 for hop2, Z23272 for hiphop1) were 5'- TTTCA TCGGC ATCAT CATCA TCATC CTT –3' (sense) and 5'- CCTTC CAGCT CCTCC ATTTC CTCTT –3' (antisense), those for VPAC1 (GenBank accession no. M86835) were 5'- GCCCC CATCC TCCTC TCCAT C –3' (sense) and 5'- TCCGC CTGCA CCTCA CCATT G –3' (antisense), and those for VPAC2 (GenBank accession no. U09631) were 5'- ATGGA TAGCA ACTCG CCTTT CTTTA G –3' (sense) and 5'- GGAAG GAACC AACAC ATAAC TCAAA CAG –3' (antisense). After initial denaturation at 94°C for 3 min, 35 cycles of amplification [30 s denaturing at 94°C, 30 s annealing at 66°C (PAC1, VPAC1) or at 63°C (VPAC2), and 1-min extension at 72°C] were performed in a DNA Thermal Cycler (Perkin Elmer, Norwalk, CT). The amplification products from RT-PCR were separated by electrophoresis (2% agarose gel in Tris-acetic acid-EDTA buffer) and visualized by ethidium bromide staining. The expected sizes of PCR products were 290 bp for the basic PAC1 receptor, 374 bp for PAC1 receptor with a single cassette insert (hip, hop1), 371 bp for PAC1 hop2 receptor, 458 bp for a double insert (hiphop1 or hiphop2), 299 bp for VPAC1, and 326 bp for VPAC2.

## 13. Measurement of Extracellular cAMP

PC12 cells ($10^5$ cells/well) in 96-well collagen I-coated plates (Becton Dickinson Labware, Bedford, MA) were stimulated for 30 min with the indicated concentrations of ligand in the serum-free medium. Supernatants were collected and aliquots (100 μl) were assayed using a kit for the determination of cyclic AMP by enzyme immunoassay according to the manufacturer's instructions (Amersham Pharmacia Biotech, Piscataway, NJ).

## 14. Statistical Analysis

For statistical comparisons, one-way analysis of variance (ANOVA) with pairwise comparison by the Fisher's least significant difference procedure was used. A $p$ value of less than 0.05 was considered significant for all analyses.

# Results and Discussion

## 1. Structural Analyses of Amylidosis-Related Peptides/proteins

The Chou-Fasman method is an empirical statistical tool for predicting peptides/proteins secondary structure using parameters obtained from the analysis of known sequences and structures [29,30], and the method was used for the prediction of amyloidogenic peptides/proteins, including PrP(106–126), Aß(1–42), and amylin (Figure 1). According to the data obtained, PrP(106–126) has two major domains, a helical conformation favoring the 109–120 sequence and a ß-sheet forming domain consisting of residues of 121–124, suggesting that PrP(106–126) has the tendency to form both helical and sheet conformation. The results indicated that Aß(1–42) has two ß-sheet forming domains at the residue of 13–19 and 31–39, and that amylin tends to form helical structure at N-terminus and turn structure at C-terminus. Upon these structural elucidations, there is no structural similarity among these amyloidogenic peptides/proteins.

| | 5 10 | 15 20 | 25 30 | 35 40 | |
|---|---|---|---|---|---|
| PrP(106–126) | K T N M K H M A G A | A A A G A V V G G L | G | | |
| *Structure* | h h h h h h h | h h h h h s s s s h | | | |
| ß-Amyloid(1–42) | D A E F R H D S G Y | E V H H Q K L V F F | A E D V G S N K G A | I I G L M V G G V V | I A |
| *Structure* | t t t t t | h h s s s s s s s h | h h t t t t t t t t | s s s s s s s s s h | s |
| Amylin | K C N T A T C A T Q | R L A N F L V H S S | N N F G A I L S S T | N V G S N T Y | |
| *Structure* | h h h | h h h h h h h t t t | t t t t s s t t t t | t t t t t t t | |

Figure 1. Sequences and secondary structure prediction of PrP(106–126), Aß(1–42), and amylin. The prediction of secondary structure was performed according to the method of Chou and Fasman. h, α-helix; s, ß-sheet; and t, turn.

CD is displayed when an optically active substance preferentially absorbs left or right-handed circularly polarized light and it provides useful information when conformational alterations occur in peptides/proteins [33]. The CD spectrum of a ß-sheet structure shows an intense positive band at 198 nm and a negative extremal band at 218 nm [34], and the CD spectrum of α-helical structure shows an intense positive peak at 192 nm and two negative peaks at 209 nm and 222 nm [33]. As shown in Figure 2, CD spectral analyses of amyloidogenic peptides/proteins, such as PrP(106–126), Aß(1–42), and amylin, indicated that they had a random coil conformation in hydrophilic condition (20 mM Tris-HCl buffer, pH 7.4), however CD spectra of PrP(106–126) and amylin exhibited α-helical structure in hydrophobic condition (50% MeOH/20 mM Tris-HCl buffer, pH 7.4). Interestingly, aging treatment of these peptides resulted in the significant conformational changes into ß-sheet structure, as evidence by the fact that PrP(106–126), Aß(1–42), and amylin gave typical CD spectra for the presence of a ß-sheet structure under both hydrophilic and hydrophobic conditions (Figure 2).

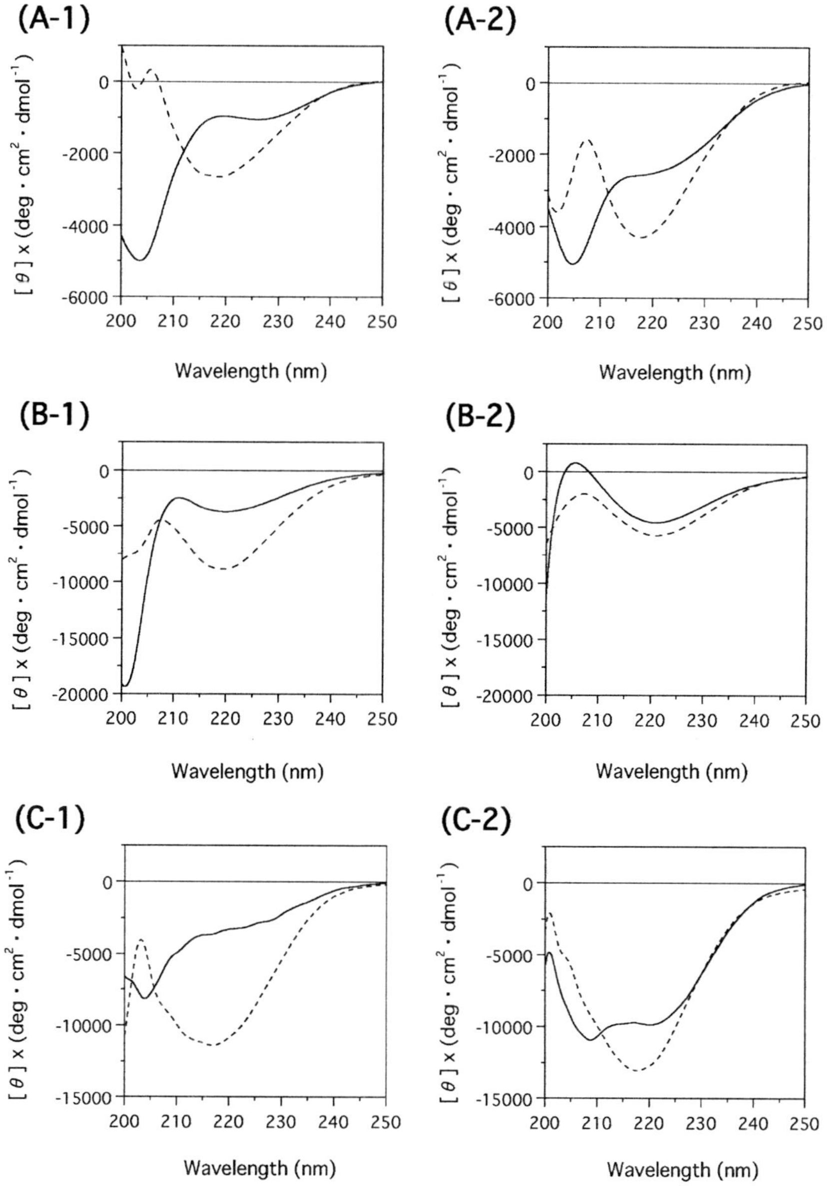


Figure 2. CD spectra representative of PrP(106–126), Aß(1–42), and amylin. (A) CD spectra of non-aged or aged PrP(106–126) in 20 mM Tris-HCl buffer, pH 7.4 (A-1) or 50% MeOH/20 mM Tris-HCl buffer, pH 7.4 (A-2). Solid line, non-aged PrP(106–126); and dashed line, PrP(106–126) aged at 10 mg/ml. (B) CD spectra of non-aged or aged Aß(1–42) in 20mM Tris-HCl buffer, pH 7.4 (B-1) or 50% MeOH/20 mM Tris-HCl buffer, pH 7.4 (B-2). Solid line, non-aged Aß(1–42); and dashed line, Aß(1–42) aged at 5 mg/ml. (C) CD spectra of non-aged or aged amylin in 20mM Tris-HCl buffer, pH 7.4 (C-1) or 50% MeOH/20 mM Tris-HCl buffer, pH 7.4 (C-2). Solid line, non-aged amylin; and dashed line, amylin aged at 5 mg/ml.

The structural changes of these amyloidogenic peptides/proteins during the aging process were supported by congo red binding assay. Generally, congo red has been used to detect amyloid fibrils in pathological specimens, since it could bind specifically to amyloid-like aggregates having a ß-sheet rich conformation [35]. Aging treatment of PrP(106–126) solution at the concentration of 10 mg/ml resulted in the significant generation of amyloid fibril as evidenced by the Congo red staining (Figure 3). In addition to PrP(106–126), Aß(1–42) and amylin, aged at the concentration of 5 mg/ml, also showed the significant generation of ß-sheet rich amyloid fibril. In addition to Congo Red, ThT has also been used commonly as an amyloid-binding dye for *in vitro* studies [36], and this fluorochrome is considered to be a potential pharmacophore for further design of amyloid-imaging agents. ThT binding activities were observed in all the tested peptides/proteins aged at high concentrations for 24 h (data not shown), and this observation was consistent with the result from congo red binding assay.

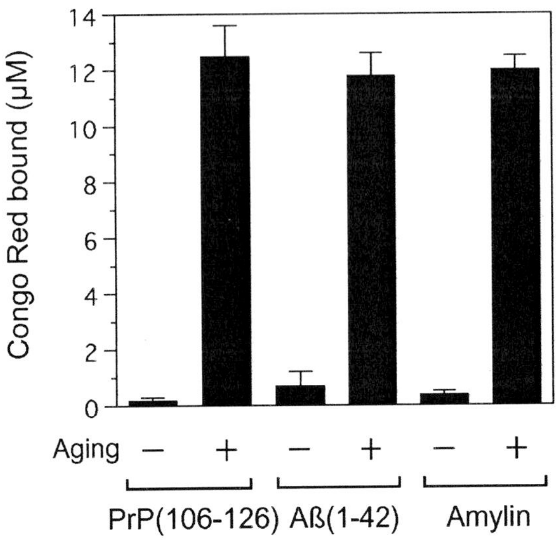


Figure 3. Aggregation behavior of PrP(106–126), Aß(1–42), and amylin as monitored by congo red binding assay. Peptides were incubated at the concentration of 10 mg/ml for PrP(106–126), and 5 mg/ml for Aß(1–42) and amylin in 0.01 M HCl at 37℃ for 24 h, and applied to congo red-binding assay.

For further clarification of amyloid fibrils, the morphologies of peptides/proteins aggregates were investigated using electron micrograph. Initially, PrP(106–126) was incubated at a concentration of 10 mg/ml in 0.01 M HCl for 24 h at 37°C and diluted to a final concentration of 1.0 mg/ml prior to applying on TEM. TEM showed well-defined fibrils (Figure 4A), and the fibril morphologies included largely disordered, rigid and branching fibrils stacked together edge to edge with a width of 10–50 nm and various lengths. Similar fibril structures were observed for Aß(1–42) and amylin aged at the concentration of 5 mg/ml for 24 h (Figure 4B and C).

Thus, the structural transition from α-helix or random coil to ß-sheet is the main conformational characteristic in the aggregation process of amyloidogenic peptides/proteins including PrP(106–126), Aß(1–42), and amylin. Despite having no sequential homology, these aggregated forms share similar ultrastructural and physicochemical properties regarding the richness of the ß-sheet conformation.

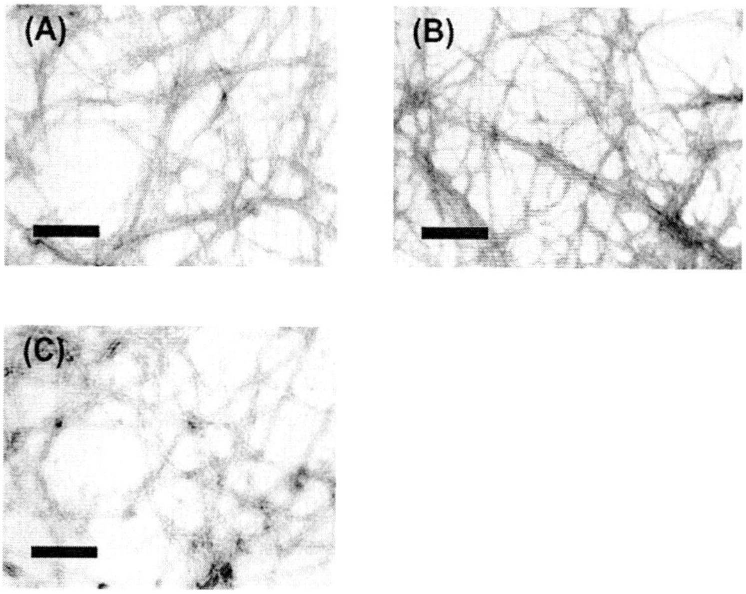


Figure 4. Electron micrographs of negatively stained fibrils formed by aged PrP(106–126) (A), Aß(1–42) (B), and amylin (C). Fibrils of PrP(106–126), Aß(1–42), and amylin were prepared by aging at the concentration of 10, 5, and 5 mg/ml, respectively, in 0.01 M HCl for 24 h. Prior to TEM analysis, the samples were diluted to the final concentration of 1 mg/ml. The bars represent 0.2 μm.

## 2. Structural Analyses of Amyloidogenic Therapeutic Peptides

In addition to amyloidosis-related peptides/proteins, some therapeutic peptides/proteins, such as glucagon and salmon calcitonin, were well known to show structural changes into amyloid-like fibrils [11]. Glucagon, consisting of 29 amino acid residues, is widely used for peroral endoscopy, clinical diacrisis and treatment of hypoglycemia, and salmon calcitonin is used for the treatment of osteoporosis and Paget's disease. The secondary structure of normal glucagon and salmon calcitonin was predicted according to the Chou-Fasman method (Figure 5). On the basis of the structural prediction, glucagon had turn structure at N-terminus and short ß-sheet structure at C-terminus. Practically, non-aged glucagon, dissolved in the hydrophilic buffer, gave the characteristic CD spectrum of a random coil conformation (Figure 6A-1). On the contrary, the CD spectra of non-aged glucagon exhibited a typical α-helical structure in the hydrophobic condition (Figure 6A-2), the content of which was estimated to be 55% according to the calculation established by Greenfield&Fasman [33]. When glucagon was aged for 24 h, a structural transition into ß-sheet structure was observed for preparations dissolved at 5 mg/ml or higher, and it resulted in a significant decrease in the α-helical content to *ca.* 1%.

With respect to the secondary structure of salmon calcitonin, it was deduced to have long helical structure in the middle part and turn-like conformation at C-terminus (Figure 5), and salmon calcitonin had no ß-sheet structure according to the Chou-Fasman method. Salmon calcitonin without aging treatment showed the presence of a random coil structure in the hydrophilic environment (Figure 6B-1) and a typical α-helical structure in the hydrophobic environment (Figure 6B-2). However, as in the case of glucagon and other amyloidogenic peptides/proteins, a conformational transition from an α-helical structure to ß-sheet structure was observed after aging at high concentrations of 50 mg/ml or higher. It was thus shown that salmon calcitonin requires much higher concentration than glucagon for structural changes to occur, suggesting that glucagon has a high propensity to yield ß-sheet structure.

| | 5 10 | 15 20 | 25 30 | |
|---|---|---|---|---|
| Glucagon (human) | H S Q G T F T S D Y | S K Y L D S R R A Q | D F V Q W L M N T | |
| *Structure* | t t t t t s t t t t | t t t s t t t t h s | s s s s s s h s s | |
| Calcitonin (salmon) | C S N L S T C V L G | K L S Q E L H K L Q | T Y P R T N T G S G | T P |
| *Structure* | t t t t s s s h h h | h h h h h h h h h h | s t t t t t t t t t | t |

Figure 5. Sequences and secondary structure prediction of glucagon and salmon calcitonin. The prediction of secondary structure was performed according to the method of Chou and Fasman. h, α-helix; s, ß-sheet; and t, turn.

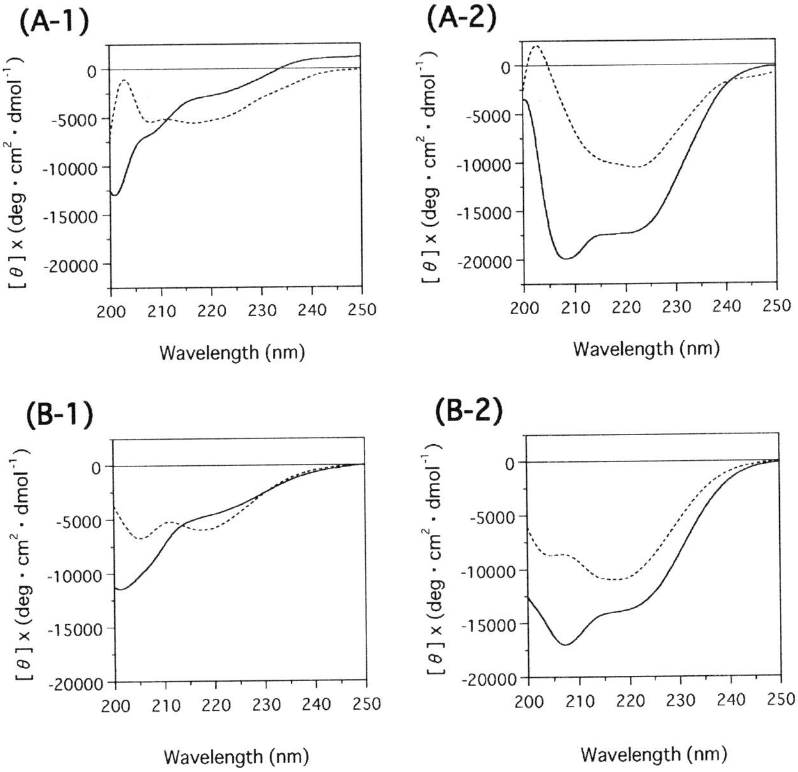


Figure 6. CD spectra representative of glucagon and salmon calcitonin. (A) CD spectra of non-aged or aged glucagon in 20 mM Tris-HCl buffer, pH 7.4 (A-1) or 50% MeOH/20 mM Tris-HCl buffer, pH 7.4 (A-2). Solid line, non-aged glucagon; and dashed line, glucagon aged at 5 mg/ml. (B) CD spectra of non-aged or aged salmon calcitoin in 20mM Tris-HCl buffer, pH 7.4 (B-1) or 50% MeOH/20 mM Tris-HCl buffer, pH 7.4 (B-2). Solid line, non-aged salmon calcitonin; and dashed line, salmon calcitoin aged at 75 mg/ml.

These observations were consistent with the results from the congo red binding assay (Figure 7). When glucagon and salmon calcitonin was incubated at the concentrations of 5 to 60 mg/ml at 37°C, bathochromic effect was observed upon staining with congo red, indicating the formation of aggregate. According to TEM image, both aged glucagon and salmon calcitonin formed well-defined fibrils (Figure 8), whose morphology resembled the aggregates of PrP(106–126) (Figure 4A) and other classic amyloid fibrils [16,37]. The fibril morphologies included largely disordered, rigid and branching fibrils stacked together edge to edge with various width and lengths.

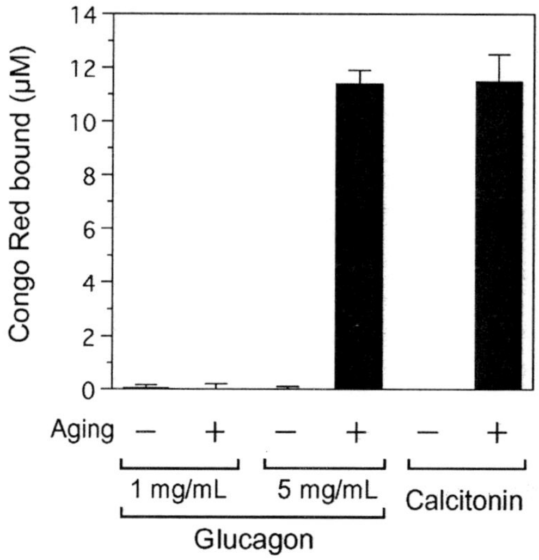


Figure 7. Aggregation behavior of glucagon and salmon calcitonin as monitored by congo red binding assay. Peptides were incubated at the concentration of 1 and 5 mg/ml for glucagon and 75 mg/ml for salmon calcitonin in 0.01 M HCl at 37℃ for 24 h, and applied to congo red-binding assay.

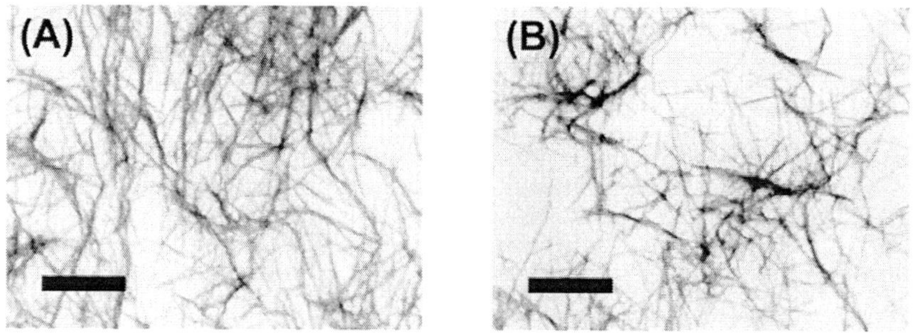


Figure 8. Electron micrographs of negatively stained fibrils formed by aged glucagon (A) and salmon calcitonin (B). Fibrils of glucagon and salmon calcitonin were prepared by aging at the concentration of 10 and 75 mg/ml, respectively, in 0.01 M HCl at 37℃ for 24 h. Prior to TEM analysis, the samples were diluted to the final concentration of 1 mg/ml. The bars represent 1.0 μm.

Next, size exclusion chromatography was conducted in order to check the size and molecular pattern of the glucagon aggregate. Under the acidic condition, the retention times for molecular weight standards and non-aged glucagon were well associated with each molecular weight (Mw) in the range from 1.2 kDa to 2,000 kDa. The correlation coefficient between retention time and Mw was estimated to be $r^2$=0.95 ($P$<0.01) according to the log(Mw) versus elution time standard curve (data not shown). Normal and aged glucagon was applied to a SEC column (ZORBAX Bio Series GF-250) at 7.5 mg/ml (Figure 9). The SEC analysis of normal glucagon, performed with 20 mM citrate-HCl buffer containing 130 mM NaCl (pH 3.0), revealed that there was only one peak at 3.5 kDa corresponding to a molecular weight of monomeric glucagon. On the other hand, a main peak appeared at 2,000 kDa in the aged glucagon, suggesting that intermolecular association is likely to be involved in the generation of the glucagon aggregate. Interestingly, the peak at 3.5 kDa of monomeric glucagon was directly shifted to a peak at 2,000 kDa without any intermediate peaks during the time course from the non-aging state to 6 h aging. Thus, SEC analysis could detect the glucagon aggregates successfully, however, the monitoring of oligomeric intermediate during aging process has been failed. In contrast to SEC analysis, capillary electrophoretic analysis of the aged glucagon could indicate the presence of some transient peaks, possibly corresponding to the oligomeric intermediates, that gradually decreased during aging treatment [38].

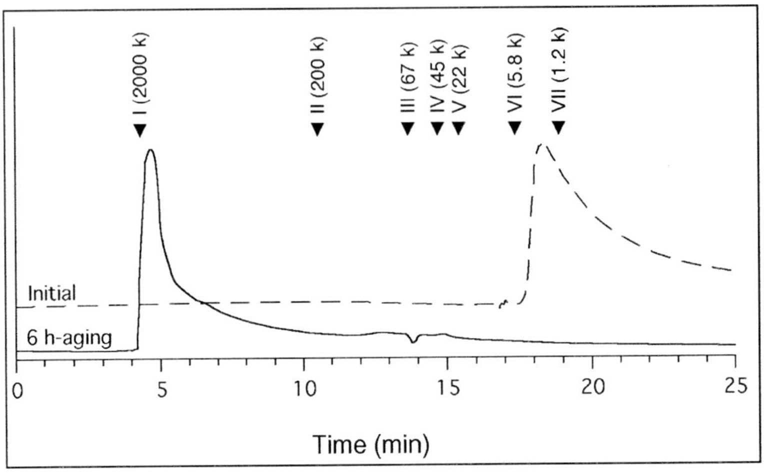


Figure 9. Size-exclusion chromatography of normal and aged glucagon. Glucagon was incubated at the concentration of 7.5 mg/ml at 37℃ for 6 h. Arrows (I–VII) represent elutions of a series of the following molecular weight protein standards with their retention times; I, blue dextran (2,000 kDa); II, Sweet potato ß-amylase (200 kDa); III, human serum albumin (66.5 kDa): IV, chicken albumin (45 kDa); V, human growth hormone (22.1 kDa); VI, porcine insulin (5.8 kDa); VII, buserelin (1.2 kDa); and *, glucagon (3.5 kDa). All samples were loaded on a SEC column (ZORBAX Bio Series GF-250) in 20 mM citrate buffer (pH 3.0), and the elution profiles were determined by UV detection at 280 nm.

In the high polar and/or acidic media, the hydrophobic side groups of amyloidogenic peptides partially tend to huddle together to avoid the aqueous environment [39]. However, under highly concentrated conditions, there are hydrophobic interactions within their respective α-helical regions to initiate aggregation. Thus, concentrated and aged glucagon molecules also seem to be self-associated to form highly ordered fibrillar aggregates with ß-sheet structures. In addition, we confirmed that aggregation of glucagon was inhibited by the addition of an anti-glucagon serum, in which the epitope was its C-terminal α-helix forming region [11]. This finding supports the concept that the α-helix region is involved in the aggregation of glucagon.

## 3. Cytotoxicity of Peptide Fibrils

To assess the cytotoxicity of the glucagon fibrils, PC12 cells, rat neuron-like cells, were exposed to 0.1–100 μM peptide aggregate for 72 h followed by the measurement of cell viability determined by WST-8 assay and released LDH. WST-8, which is reduced by mitochondria of living cells, is widely used to measure cell viability as an alternative to MTT. A significant decrease in cell viability was observed in cultures exposed to 10–100 μM aged glucagon or PrP(106–126) ($P<0.01$) but not in cultures treated with 100 μM non-aged peptides (Figure 10). Toxicity of both peptide fibrils seemed to be concentration-dependent, and the aged PrP(106–126) showed much higher toxicity in PC12 cells as compared to aged glucagon.

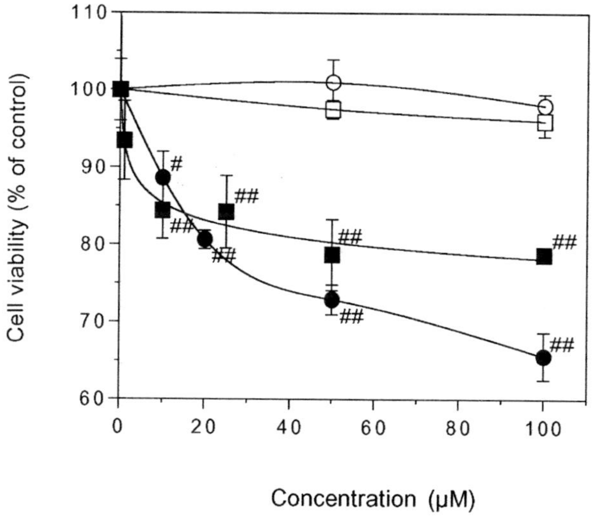


Figure 10. Effects of aged PrP(106–126) and glucagon on cell viability. PC12 cells were treated with different concentrations of non-aged or aged peptides for 72 h, then the cell viability was determined using WST-8 assay. PrP(106–126), ○; aged PrP(106–126), ●; glucagon, □; and aged glucagon, ■. Data represent the mean±SD of six determinations. ##, $p<0.01$ and #, $p<0.05$ with respect to the control group (vehicle alone).

To determine whether the loss of cell viability was equivalent to cell death, the release of LDH was measured. Treatment with 10 μM aged glucagon induced a significant increase in LDH release compared to control (Figure 11) and no significant increase in LDH release was observed in cultures treated with non-aged glucagon at 100 μM and aged glucagon at 1μM or lower (data not shown). PrP(106–126) also stimulated the gradual release of cellular LDH into the culture medium, indicating the loss of cell membrane integrity in both necrotic and apoptotic cells. The increase in LDH release by PrP(106–126) was evident at 48 h after the treatment although the control medium treatment led to a slight induction of the release of LDH.

Thus, glucagon fibrils were found to be highly toxic to PC12 cells, which was similar to the case of aged PrP(106–126) and Aß(1-42) (>10 μM) [40,41]. Aged salmon calcitonin also displayed significant cytotoxicity in PC12 cells (data not shown), whereas non-aged salmon calcitonin did not induce significant cell death. These findings suggested that the presence of amyloid fibrils was responsible for the cytotoxicity of aged glucagon, aged PrP(106–126) or other aged peptides/proteins toward neuronal cells. These results were consistent with a previous report showing that peptide fibrils induced neurotoxicity in cultured hippocampal neurons while native peptide did not affect the cell survival as measured by MTT assay [42].

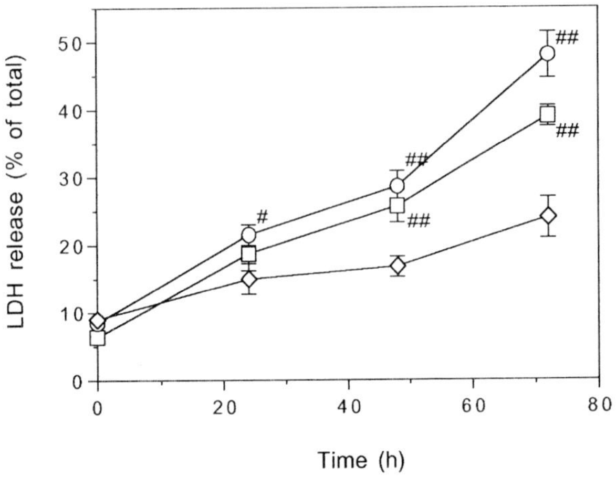


Figure 11. Neurotoxic effects of aged PrP(106–126) and glucagon in PC12 cells. PC12 cells were treated with aged PrP(106–126) or glucagon(50 μM) for indicated periods, and the medium was assayed by the release of LDH. Control (vehicle alone), ◇; aged PrP(106–126), ○; and aged glucagon, □. The results are the means±SD of four experiments. $^{\#\#}$, $p$<0.01 and $^{\#}$, $p$<0.05 with respect to the control group.

## 4. Signaling Pathways in the Cytotoxicity of Peptide Fibrils

The family of caspase enzymes is a large group of proteases whose members have defined roles in apoptotic cell death. In particular, the activation of caspase-3 is required for the early stages of apoptosis that include DNA fragmentation and morphological changes [43], and the apoptosis is induced by many neurotoxic agents including nitric oxide [44], tumor necrosis factor-α, and cycloheximide [45]. To determine whether aged peptides/proteins induces caspase-3 activation in PC12 cells, we exposed PC12 cells to the aged glucagon or PrP(106–126) (50 μM) and measured caspase-3-like activity in cell lysates obtained by the cleavage of a fluorometric caspase-3 substrate, Z-DEVD-Rhodamine 110 [42]. Exposure of PC12 cell to aged glucagon resulted in the significant increase of the caspase-3 activity just prior to the loss of membrane integrity, and 24 h after incubation the maximal caspase-3 activity was detected at 160% of the control level (Figure 12). On the contrary, no significant elevation of caspase-3 activity was observed in cells treated with non-aged glucagon (50 μM). With respect to aged PrP(106–126)-treated cells, the activation of caspase-3 reached the maximal level at 12 h, and this elevated caspase-3 activity decreased gradually to the same level as control at 48 h. These data indicated that the exposure of PC12 cells to the peptide fibrils induced a rapid and significant elevation in caspase-3 activity within 24 h prior to the loss of cell viability that occurred 72 h after exposure.

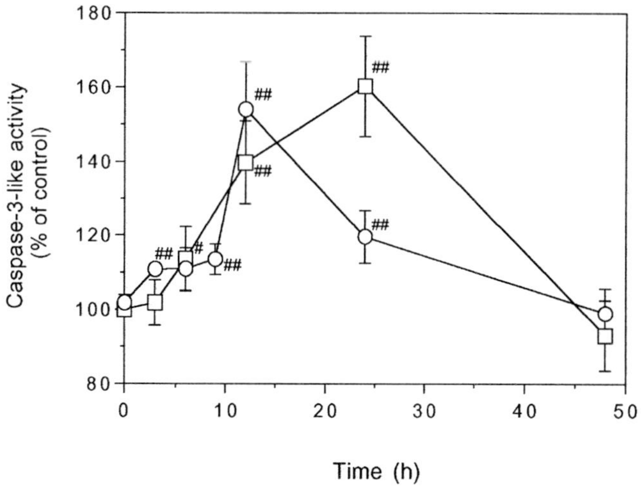


Figure 12. Time course of caspase-3 activity in cytosolic protein extracts from PC12 cells treated with aged PrP(106–126) and glucagon. PC12 cells were exposed to aged PrP(106–126) or glucagon (50 μM) and lysed for the indicated period. Caspase-3-like protease activity was determined by the cleavage of the fluorometrical caspase-3 substrate, Z-DEVD-Rhodamine 110. Aged PrP(106–126), ○; and aged glucagon, □. The data are expressed as percentage of the control value (means±SD of four experiments). ##, $p<0.01$ and #, $p<0.05$ with respect to the control group.

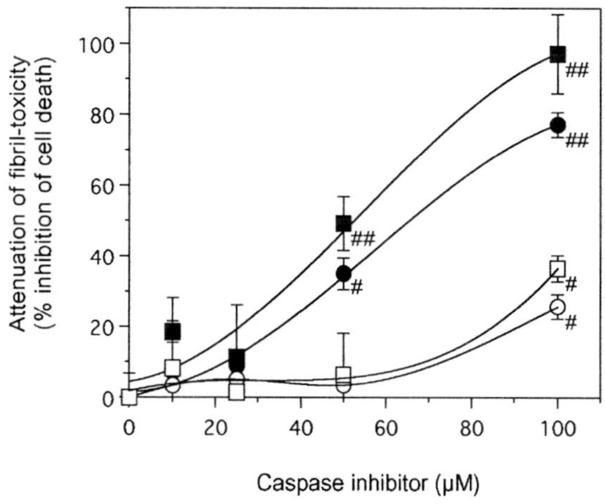


Figure 13. Effects of caspase inhibitors on the cytotoxicity induced by aged PrP(106–126) and glucagon. PC12 cells were chronically treated with 50 μM of aged PrP(106–126) with Ac-DEVD-CHO (○) or Z-VAD-FMK (●), or 50 μM of aged glucagon with Ac-DEVD-CHO (□) or Z-VAD-FMK (■) at various concentrations for 72 h. Cell viability was determined by WST-8 assay. Each value is the means±SD of four experiments. ##, $p<0.01$ and #, $p<0.05$ with respect to the control group (caspase inhibitor free).

Inhibitors of caspases including Ac-DEVD-CHO, a caspase-3-specific inhibitor [46], and Z-VAD-FMK, an irreversible inhibitor of several members of the caspase family [47], were employed to investigate whether apoptosis was involved in the cytotoxicity by aged glucagon or PrP(106–126). The addition of Z-VAD-FMK (100 μM) rescued both glucagon and PrP(106–126)-induced cell death close to the control level (vehicle only) and this protective effect was reduced to half at 50 μM (Figure 13). On the other hand, Ac-DEVD-CHO attenuated the fibril toxicity by 30–40% only at high concentration (100 μM). These data indicate that not only caspase-3 but also other caspases may play an important role in the final execution of the cell death program stimulated by peptide fibrils. These results, taken together with the physicochemical properties of amyloidogenic peptides/proteins, provides further evidence for the concept that there is an intrinsic cytotoxic signaling pathway in fibrillar aggregates from various peptides/proteins, suggesting that this cytotoxic mechanism occurs in a conformation-specific manner rather than a sequence-specific manner.

## 5. Expression of the mRNA for PACAP/VIP Receptors in PC12 Cells

There are three types of receptors recognized by the PACAP/VIP family, which include "PACAP-preferring" PAC1, "VIP-preferring" VPAC1, and "helodermine-preferring" VPAC2 [48]. PACAP and VIP have approximately equal affinity for the VPAC1 and VPAC2 receptors; however, PACAP has about 1000 times higher binding activity than VIP toward PAC1 receptors. PACAP/VIP receptors are widely distributed glycoproteins with a large hydrophilic extracellular domain followed by seven highly conserved hydrophobic transmembrane helices [23]. With respect to the tissue distribution of these receptors, the PAC1 receptor is mainly localized in the central nervous system and the anterior pituitary, while the VPAC receptors are generally concentrated in the peripheral tissues such as the lung, liver and kidney [49]. Although the expression of PAC1 receptor mRNA in the central nervous systems was found to be much higher than those of VPAC1 and VPAC2 receptors mRNA [50], expression of VPAC2 receptor has been confirmed in some brain tissues including hypothalamus and hippocampus [51-53].

In this investigation, we demonstrated the effect of neuropeptides on the fibril-induced neuronal damage using neuron-like PC12 cells. PC12 cells are typical paraneuronal cultured cells that originated from a rat adrenal pheochromocytoma, and some neuroactive substances, including nerve growth factor, fibroblast growth factors (acidic and basic FGF), cAMP, cell adhesive substances (laminin and NCAM), manganese, staurosporine, FK 506, were found to elicit notable inductions of neurite outgrowth in PC-12 cells [26]. To investigate the expression level of PACAP/VIP receptors in PC12 cells, RT-PCR was performed. As shown in Figure 14, a single band of the predicted size for the PAC1 receptor was amplified. In contrast, no products were detected for VPAC1 or VPAC2 receptors, whereas they were amplified from rat aorta mRNA [54,55]. In parallel control experiments without reverse transcription, PCR products for the ß-actin and PACAP/VIP receptors were barely detectable, indicating that the amplified PAC1 receptor product was not derived from contaminating genomic or mitochondrial DNA. Thus, PAC1 receptor is dominantly expressed in PC12 cells.

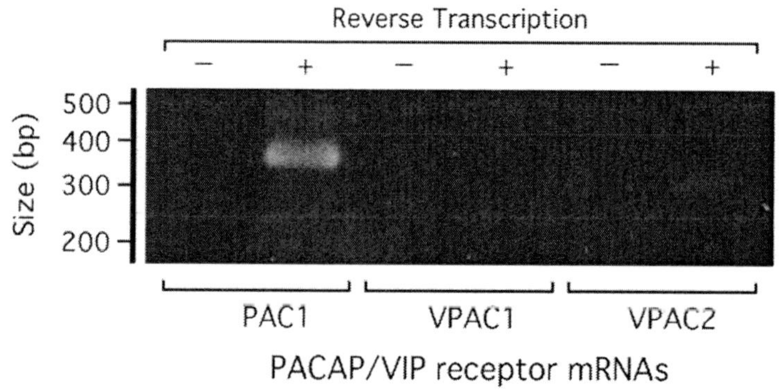


Figure 14. RT-PCR analysis of PACAP/VIP receptor mRNAs in PC12 cells. Total RNAs were reverse transcribed in the presence (RT+) or absence (RT–) of reverse transcriptase and amplified by PCR with oligonucleotide primers specific for each of the PACAP/VIP receptors.

PACAP was discovered as a substance that increased cAMP level [17]. There are two forms of mammalian PACAP, PACAP38 and PACAP27, and they have been shown to have the same stimulatory effect on cAMP production [18,56]. As shown in Figure 15, PACAP27 increased cAMP formation in PC12 cells, whereas VIP was less potent than PACAP27 in stimulating cAMP formation. This was consistent with the results of the RT-PCR experiment, which showed the predominant expression of the PAC1 receptor among the PACAP/VIP family receptors in PC12 cells.

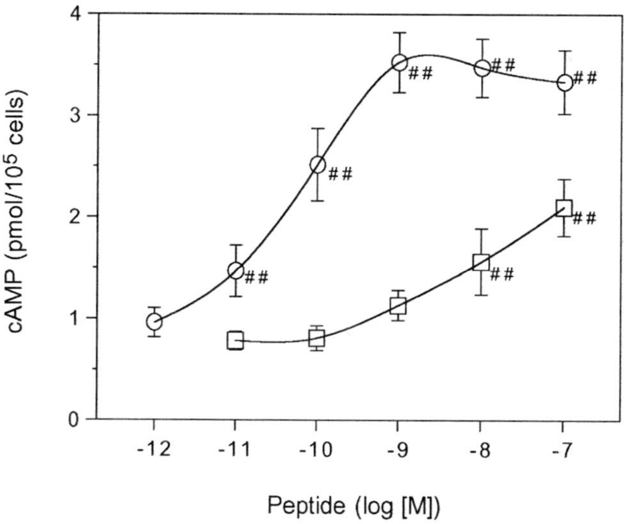


Figure 15. Accumulation of extracellular cAMP stimulated by neuropeptides. PC12 cells, plated at a density of $10^5$ cells/well in 96-well plates coated with type I collagen, were stimulated with PACAP27 (○) or VIP (□) for 30 min, and extracellular cAMP levels were determined by enzyme immunoassay. The data represent the averages of four determinations and are shown as the means±SD. $^{\#\#}$, $p<0.01$ with respect to the control group (vehicle alone).

## 6. The Protective Effects of PACAP27 on PrP(106–126)-Induced Neurocytotoxicity

Although chronic exposure of PC12 cells to PrP(106–126) resulted in a concentration-dependent decrease of cell viability assayed by WST-8, PACAP27 suppressed PrP(106–126)-induced cell death at concentrations as low as $10^{-15}$ M (Figure 16). Interestingly, the cytoprotective effect of PACAP27 against PrP(106–126)-induced toxicity was diminished at $10^{-7}$ M, providing a bell-shaped dose-dependency. Incubation of PC12 cells with VIP ($10^{-7}$ M) also attenuated the cytotoxicity of PrP(106–126), and these dose-response fashions were consistent with the data of the PACAP27-evoked activation of adenylate cyclase. The present study revealed that PACAP27 was $10^4$-fold higher than VIP in neuroprotection against PrP(106–126)-cytotoxicity. In addition, PACAP was much higher than VIP in activation of adenylate cyclase, catecholamine secretion, and neurite outgrowth in PC12 cells [26,27]. PACAP27 and VIP display similar affinities for VPAC1 and 2 receptors, whereas the binding activity of PACAP toward PAC1 receptors is $10^3$–$10^4$ fold higher than VIP. These results, taken together with the results showing the predominant expression of PAC1 receptor in PC12 cells, indicated that the effect of PACAP/VIP is mediated through the activation of "PACAP-preferring" PAC1 receptors in PC12 cells. Hence, this is the reason why PACAP27 at the extremely low concentration shows significant biological activities. It was also demonstrated that PACAP protected neuronal cells against apoptosis induced by different neurotoxins, such as ethanol [57], free radicals [58], glutamate receptor agonists [59,60], and gp120, an envelope glycoprotein from the human immunodeficiency virus (HIV) [61]. The present study, taken together with these previous outcomes, provided further evidence for the neuroprotective potential of PACAP, as well as VIP, in the neuronal cells and nervous tissues.

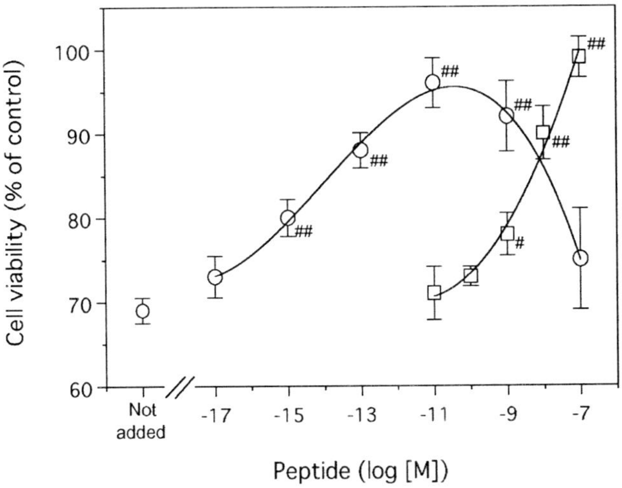


Figure 16. Protective effects of neuropeptides against aged PrP(106–126)-induced cell death of PC12 cells. PACAP27 (○) or VIP (□) was added into the culture medium at a final concentration between $10^{-17}$ and $10^{-7}$ M, and the cells were incubated with aggregated PrP(106–126) (50 μM). After a 72-h incubation, cell viability was assessed by measuring the reduction of WST-8 (means±SE of four experiments). $^{\#\#}$, $p$<0.01 and $^{\#}$, $p$<0.05 with respect to the control group (vehicle alone).

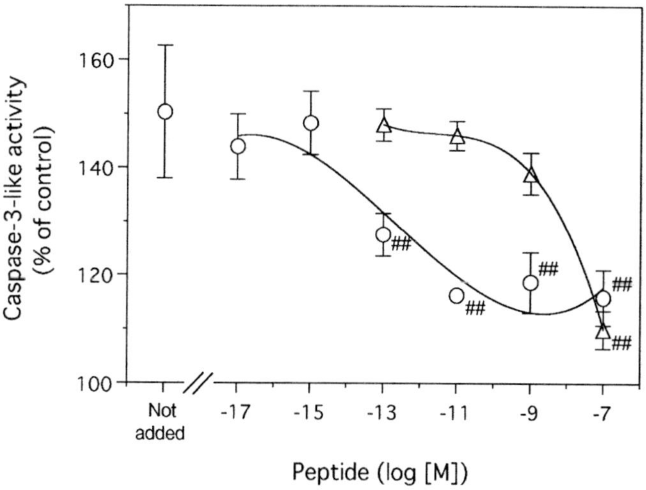


Figure 17. Effects of neuropeptides on PrP(106–126)-evoked caspase-3 activity. PC12 cells were exposed for 12 h to PrP(106–126) (50 μM) with PACAP27 (○) or VIP (□), followed by measurement of the caspase-3-like activity. Each value is the means±SD of four experiments. ##, $p<0.01$ with respect to the PrP(106–126)-treated group.

## 7. Effect of PACAP27 on Caspase-3-like Activity

Apoptosis is a physiologically important cellular suicide pathway, and there is some evidence to indicate that the mechanism of neuronal cell death in prion disease is apoptosis as apoptotic neurons have been observed in the brain of scrapie-infected sheep [62]. Apoptosis is mediated by the family of cysteine, aspartyl-specific proteases known as caspases, and the exposure of PC12 cells to PrP(106–126) fibril resulted in the significant activation of caspase-3, reaching the maximal level at 12 h. Then, PACAP27, at the concentrations between $10^{-13}$ and $10^{-7}$ M, significantly deactivated the PrP(106–126)-evoked caspase-3 activity up to 70% inhibition, producing a dose-dependent curve (Figure 17). The addition of PACAP27 resulted in the significant decrease of cell death and catalytic activity of caspase-3, suggesting that PACAP27 acted as an anti-apoptotic agent through the deactivation of caspase-3. VIP, as well as PACAP27, at concentrations of $10^{-7}$ M also attenuated the PrP(106–126)-induced stimulation of caspase-3 activity in PC12 cells.

## 8. Signaling Pathways Involved in the Neuroprotective Effect of PACAP27

PACAP stimulates various protein kinases including the adenylate cyclase/PKA, the phospholipase C/PKC and the mitogen-activated protein (MAP) kinase pathways [23]. PACAP exerts its neuronal actions through the activation of these kinases in the nervous system. In order to clarify the possible signaling pathway involved in the neuroprotective effect of PACAP27, the effect of stimulators/inhibitors of protein kinases were examined. db-cAMP, a potent PKA stimulator [63], showed a dose-dependent neuroprotection against the toxicity of PrP(106–126) (data not shown). A potent PKC activator, 12-*O*-tetradecanoyl-phorbol-13-acetate (TPA; $10^{-7}$ M), [64], reduced the cytotoxicity of PrP(106–126), whereas TPA at the higher concentrations (>$10^{-6}$ M) showed no neuroprotection due to the cytotoxicity itself (data not shown). Next, when the selective PKA inhibitor, H89, was added simultaneously with PrP(106–126) and PACAP27 ($10^{-9}$ M), the protective effect of PACAP27 against PrP(106–126)-induced neurotoxicity was inhibited significantly (Figure 18). U0126, a specific MAP kinase kinase inhibitor [65], caused a marked attenuation of the PACAP27-evoked neuroprotection. On the other hand, the selective PKC inhibitor, Myr-ψPKC ($10^{-6}$ M) [66], showed no significant response against the effect of PACAP27 in this manipulation. Hence, the neuroprotection of PACAP27 was mediated through PKA and MAP kinase, but not through PKC signaling pathways.

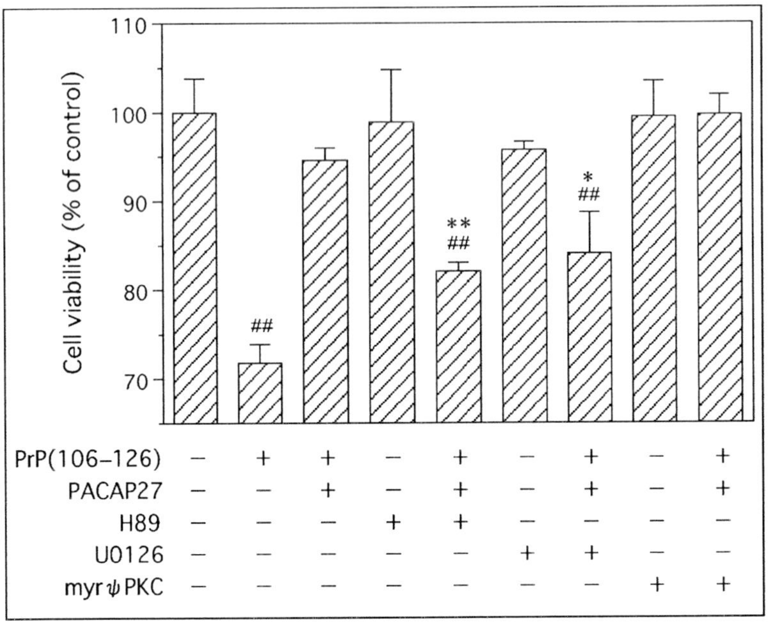


Figure 18. Involvement of protein kinase cascades in the neuroprotective effect of PACAP27. PC12 cells were treated with PrP(106–126) (50 μM) and PACAP27 ($10^{-9}$ M) in the absence or presence of the protein kinase inhibitors ($10^{-6}$ M) for 72 h, and the cell viability was assessed by WST-8 reduction analysis. The data represent the means±SE of four experiments. ##, *p*<0.01 with respect to the control group. **, *p*<0.01 and *, *p*<0.05 between PACAP27-treated group with and without protein kinase inhibitor.

Previously, we demonstrated that cigarette smoke extracts (CSE) induced the apoptotic death of rat alveolar L2 cells, cloned from adult rat alveolar epithelial cells [54,67,68]. RT-PCR experiment revealed the exclusive expression of VPAC2 receptors in L2 cells, and this is consistent with our previous study which showed the predominant expression of the VPAC2 receptor in rat lung [69]. Both VIP and PACAP27 at extremely low concentrations effectively attenuated the decrease in the viability of L2 cells induced by CSE, in a concentration-dependent manner [54,55]. The results suggested that the protective effect of VIP on the CSE-induced cell death in L2 cells was mediated through stimulation of the VPAC2 receptor, followed possibly by the activation of the PKA and MAP kinase signaling pathways. Furthermore, the data obtained indicated that the deactivation of CSE-stimulated matrix metalloproteinases and caspase-3 was one of the possible mechanisms for the protective effect of VIP on CSE-induced cytotoxicity in L2 cells. These observations were very similar to the results from the present investigation, so VIP, as well as PACAP, would be effective in the neuroprotection against fibril-induced toxicity in some neuronal cells and tissues, where the VPAC1 and VPAC2 receptors are expressed.

## CONCLUSION

In this investigation, we demonstrated that the aging treatment of amyloidogenic peptides/proteins, including PrP(106–126), Aß(1–42), amylin, glucagon, and salmon calcitonin, resulted in the significant generation of amyloidogenic fibrils, leading to cytotoxicity mediated by activation of the apoptotic enzyme caspase-3 *in vitro*. We showed the neuroprotective effect of PACAP27 on PrP(106–126)-induced neuronal cell death, and the neuroprotection of PACAP27 was significantly effective at extremely low concentrations ($10^{-15}$ M). The PACAP-related neuropeptide, VIP, also showed a similar neuroprotective effect only at the higher concentrations (>$10^{-9}$ M). A caspase-3-specific inhibitor, Ac-DEVD-CHO, slightly deactivated the caspase-3 in PrP(106-126)-stimulated PC12 cells to increase the cell viability, while the broad-spectrum caspase inhibitor, Z-VAD-FMK, showed both clear prevention of caspase-activation and cell death. These results suggest that there is activation of critical cytotoxic factors upstream of caspase activation, and that PACAP27 may inhibits the neurotoxicity of PrP(106-126) through caspase-3-dependent and independent pathways. PACAP27 activates both the PKA and MAP kinase signaling pathways through the PAC1 receptor, which may lead to the rescue of apoptotic cell death induced by PrP and other peptides/proteins fibrils.

## REFERENCES


[1] Carrell, R. W., and Lomas, D. A. Conformational disease. *Lancet* (1997) 350: 134-138.

[2] Glenner, G. G. Amyloid deposits and amyloidosis. The beta-fibrilloses (first of two parts). *N Engl J Med* (1980) 302: 1283-1292.

[3] Glenner, G. G. Amyloid deposits and amyloidosis: the beta-fibrilloses (second of two parts). *N Engl J Med* (1980) 302: 1333-1343.

[4] Stefani, M., and Dobson, C. M. Protein aggregation and aggregate toxicity: new insights into protein folding, misfolding diseases and biological evolution. *J Mol Med* (2003) 81: 678-699.

[5] Ross, C. A., and Poirier, M. A. Protein aggregation and neurodegenerative disease. *Nat Med* (2004) 10 Suppl: S10-17.

[6] Xing, Y., and Higuchi, K. Amyloid fibril proteins. *Mech Ageing Dev* (2002) 123: 1625-1636.

[7] Vines, G. Alzheimer's disease--from cause to cure? *Trends Biotechnol* (1993) 11: 49-55.

[8] Cooper, G. J., Willis, A. C., Clark, A., Turner, R. C., Sim, R. B., and Reid, K. B. Purification and characterization of a peptide from amyloid-rich pancreases of type 2 diabetic patients. *Proc Natl Acad Sci U S A* (1987) 84: 8628-8632.

[9] Prusiner, S. B. Prions. *Proc Natl Acad Sci U S A* (1998) 95: 13363-13383.

[10] Scherzinger, E., Lurz, R., Turmaine, M., Mangiarini, L., Hollenbach, B., Hasenbank, R., Bates, G. P., Davies, S. W., Lehrach, H., and Wanker, E. E. Huntingtin-encoded polyglutamine expansions form amyloid-like protein aggregates in vitro and in vivo. *Cell* (1997) 90: 549-558.

[11] Onoue, S., Ohshima, K., Debari, K., Koh, K., Shioda, S., Iwasa, S., Kashimoto, K., and Yajima, T. Mishandling of the therapeutic peptide glucagon generates cytotoxic amyloidogenic fibrils. *Pharm Res* (2004) 21: 1274-1283.

[12] Ellis, R. J., and Pinheiro, T. J. Medicine: danger--misfolding proteins. *Nature* (2002) 416: 483-484.

[13] Sipe, J. D. Amyloidosis. *Annu Rev Biochem* (1992) 61: 947-975.

[14] Turnell, W. G., and Finch, J. T. Binding of the dye congo red to the amyloid protein pig insulin reveals a novel homology amongst amyloid-forming peptide sequences. *J Mol Biol* (1992) 227: 1205-1223.

[15] Kurochkin, I. V. Amyloidogenic determinant as a substrate recognition motif of insulin-degrading enzyme. *FEBS Lett* (1998) 427: 153-156.

[16] Forloni, G., Angeretti, N., Chiesa, R., Monzani, E., Salmona, M., Bugiani, O., and Tagliavini, F. Neurotoxicity of a prion protein fragment. *Nature* (1993) 362: 543-546.

[17] Miyata, A., Arimura, A., Dahl, R. R., Minamino, N., Uehara, A., Jiang, L., Culler, M. D., and Coy, D. H. Isolation of a novel 38 residue-hypothalamic polypeptide which stimulates adenylate cyclase in pituitary cells. *Biochem Biophys Res Commun* (1989) 164: 567-574.

[18] Miyata, A., Jiang, L., Dahl, R. D., Kitada, C., Kubo, K., Fujino, M., Minamino, N., and Arimura, A. Isolation of a neuropeptide corresponding to the N-terminal 27 residues of the pituitary adenylate cyclase activating polypeptide with 38 residues (PACAP38). *Biochem Biophys Res Commun* (1990) 170: 643-648.

[19] Arimura, A. Pituitary adenylate cyclase activating polypeptide (PACAP): discovery and current status of research. *Regul Pept* (1992) 37: 287-303.

[20] Said, S. I., and Mutt, V. Polypeptide with broad biological activity: isolation from the small intestine. *Science* (1970) 169: 1217-1218.

[21] Said, S. I. Vasoactive intestinal polypeptide (VIP) in asthma. *Ann N Y Acad Sci* (1991) 629: 305-318.

[22] Delgado, M., Munoz-Elias, E. J., Gomariz, R. P., and Ganea, D. Vasoactive intestinal peptide and pituitary adenylate cyclase-activating polypeptide enhance IL-10 production by murine macrophages: in vitro and in vivo studies. *J Immunol* (1999) 162: 1707-1716.

[23] Vaudry, D., Gonzalez, B. J., Basille, M., Yon, L., Fournier, A., and Vaudry, H. Pituitary adenylate cyclase-activating polypeptide and its receptors: from structure to functions. *Pharmacol Rev* (2000) 52: 269-324.

[24] Shioda, S., Ozawa, H., Dohi, K., Mizushima, H., Matsumoto, K., Nakajo, S., Takaki, A., Zhou, C. J., Nakai, Y., and Arimura, A. PACAP protects hippocampal neurons

against apoptosis: involvement of JNK/SAPK signaling pathway. *Ann N Y Acad Sci* (1998) 865: 111-117.

[25] Shoge, K., Mishima, H. K., Saitoh, T., Ishihara, K., Tamura, Y., Shiomi, H., and Sasa, M. Protective effects of vasoactive intestinal peptide against delayed glutamate neurotoxicity in cultured retina. *Brain Res* (1998) 809: 127-136.

[26] Onoue, S., Waki, Y., Nagano, Y., Satoh, S., and Kashimoto, K. The neuromodulatory effects of VIP/PACAP on PC-12 cells are associated with their N-terminal structures. *Peptides* (2001) 22: 867-872.

[27] Onoue, S., Waki, Y., Hamanaka, K., Takehiko, Y., and Kashimoto, K. Vasoactive intestinal peptide regulates catecholamine secretion in rat PC12 cells through the pituitary adenylate cyclase activating polypeptide receptor. *Biomed Res* (2001) 22: 77-82.

[28] Merrifield, R. B. Solid-phase peptide synthesis. *Adv Enzymol Relat Areas Mol Biol* (1969) 32: 221-296.

[29] Chou, P. Y., and Fasman, G. D. Empirical predictions of protein conformation. *Annu Rev Biochem* (1978) 47: 251-276.

[30] Chou, P. Y., and Fasman, G. D. Prediction of protein conformation. *Biochemistry* (1974) 13: 222-245.

[31] Klunk, W. E., Pettegrew, J. W., and Abraham, D. J. Two simple methods for quantifying low-affinity dye-substrate binding. *J Histochem Cytochem (*1989) 37: 1293-1297.

[32] Isobe, I., Michikawa, M., and Yanagisawa, K. Enhancement of MTT, a tetrazolium salt, exocytosis by amyloid beta- protein and chloroquine in cultured rat astrocytes. *Neurosci Lett* (1999) 266: 129-132.

[33] Greenfield, N., and Fasman, G. D. Computed circular dichroism spectra for the evaluation of protein conformation. *Biochemistry* (1969) 8: 4108-4116.

[34] Cort, J., Liu, Z., Lee, G., Harris, S. M., Prickett, K. S., Gaeta, L. S., and Andersen, N. H. Beta-structure in human amylin and two designer beta-peptides: CD and NMR spectroscopic comparisons suggest soluble beta-oligomers and the absence of significant populations of beta-strand dimers. *Biochem Biophys Res Commun* (1994) 204: 1088-1095.

[35] Klunk, W. E., Pettegrew, J. W., and Abraham, D. J. Quantitative evaluation of congo red binding to amyloid-like proteins with a beta-pleated sheet conformation. *J Histochem Cytochem* (1989) 37: 1273-1281.

[36] LeVine, H., 3rd. Thioflavine T interaction with synthetic Alzheimer's disease beta-amyloid peptides: detection of amyloid aggregation in solution. *Protein Sci* (1993) 2: 404-410.

[37] Solomon, B., Koppel, R., Hanan, E., and Katzav, T. Monoclonal antibodies inhibit in vitro fibrillar aggregation of the Alzheimer beta-amyloid peptide. *Proc Natl Acad Sci U S A* (1996) 93: 452-455.

[38] Onoue, S., Iwasa, S., Kojima, T., Katoh, F., Debari, K., Koh, K., Matsuda, Y., and Yajima, T. Structural transition of glucagon in the concentrated solution observed by electrophoretic and spectroscopic techniques. *J Chromatogr A* (2006) in press.

[39] Sasaki, K., Dockerill, S., Adamiak, D. A., Tickle, I. J., and Blundell, T. X-ray analysis of glucagon and its relationship to receptor binding. *Nature* (1975) 257: 751-757.

[40] Onoue, S., Ohshima, K., Endo, K., Yajima, T., and Kashimoto, K. PACAP protects neuronal PC12 cells from the cytotoxicity of human prion protein fragment 106-126. *FEBS Lett* (2002) 522: 65-70.

[41] Onoue, S., Endo, K., Ohshima, K., Yajima, T., and Kashimoto, K. The neuropeptide PACAP attenuates beta-amyloid (1-42)-induced toxicity in PC12 cells. *Peptides* (2002) 23: 1471-1478.

[42] Liu, J., Bhalgat, M., Zhang, C., Diwu, Z., Hoyland, B., and Klaubert, D. H. Fluorescent molecular probes V: a sensitive caspase-3 substrate for fluorometric assays. *Bioorg Med Chem Lett* (1999) 9: 3231-3236.

[43] Janicke, R. U., Sprengart, M. L., Wati, M. R., and Porter, A. G. Caspase-3 is required for DNA fragmentation and morphological changes associated with apoptosis. *J Biol Chem* (1998) 273: 9357-9360.

[44] Chae, H. J., Chae, S. W., An, N. H., Kim, J. H., Kim, C. W., Yoo, S. K., Kim, H. H., Lee, Z. H., and Kim, H. R. Cyclic-AMP inhibits nitric oxide-induced apoptosis in human osteoblast: the regulation of caspase-3, -6, -9 and the release of cytochrome c in nitric oxide-induced apoptosis by cAMP. *Biol Pharm Bull* (2001) 24: 453-460.

[45] Li, J., Yang, S., and Billiar, T. R. Cyclic nucleotides suppress tumor necrosis factor alpha-mediated apoptosis by inhibiting caspase activation and cytochrome c release in primary hepatocytes via a mechanism independent of Akt activation. *J Biol Chem* (2000) 275: 13026-13034.

[46] Nicholson, D. W., Ali, A., Thornberry, N. A., Vaillancourt, J. P., Ding, C. K., Gallant, M., Gareau, Y., Griffin, P. R., Labelle, M., Lazebnik, Y. A., and et al. Identification and inhibition of the ICE/CED-3 protease necessary for mammalian apoptosis. *Nature* (1995) 376: 37-43.

[47] Atkinson, E. A., Barry, M., Darmon, A. J., Shostak, I., Turner, P. C., Moyer, R. W., and Bleackley, R. C. Cytotoxic T lymphocyte-assisted suicide. Caspase 3 activation is primarily the result of the direct action of granzyme B. *J Biol Chem* (1998) 273: 21261-21266.

[48] Rawlings, S. R., and Hezareh, M. Pituitary adenylate cyclase-activating polypeptide (PACAP) and PACAP/vasoactive intestinal polypeptide receptors: actions on the anterior pituitary gland. *Endocr Rev (*1996) 17: 4-29.

[49] Shivers, B. D., Gorcs, T. J., Gottschall, P. E., and Arimura, A. Two high affinity binding sites for pituitary adenylate cyclase- activating polypeptide have different tissue distributions. *Endocrinology* (1991) 128: 3055-3065.

[50] Basille, M., Vaudry, D., Coulouarn, Y., Jegou, S., Lihrmann, I., Fournier, A., Vaudry, H., and Gonzalez, B. Comparative distribution of pituitary adenylate cyclase-activating polypeptide (PACAP) binding sites and PACAP receptor mRNAs in the rat brain during development. *J Comp Neurol* (2000) 425: 495-509.

[51] Hashimoto, H., Ishihara, T., Shigemoto, R., Mori, K., and Nagata, S. Molecular cloning and tissue distribution of a receptor for pituitary adenylate cyclase-activating polypeptide. *Neuron* (1993) 11: 333-342.

[52] Usdin, T. B., Bonner, T. I., and Mezey, E. Two receptors for vasoactive intestinal polypeptide with similar specificity and complementary distributions. *Endocrinology* (1994) 135: 2662-2680.

[53] Shioda, S., Shuto, Y., Somogyvari-Vigh, A., Legradi, G., Onda, H., Coy, D. H., Nakajo, S., and Arimura, A. Localization and gene expression of the receptor for pituitary adenylate cyclase-activating polypeptide in the rat brain. *Neurosci Res* (1997) 28: 345-354.

[54] Onoue, S., Ohmori, Y., Endo, K., Yamada, S., Kimura, R., and Yajima, T. Vasoactive intestinal peptide and pituitary adenylate cyclase-activating polypeptide attenuate the cigarette smoke extract-induced apoptotic death of rat alveolar L2 cells. *Eur J Biochem* (2004) 271: 1757-1767.

[55] Onoue, S., Endo, K., Ohmori, Y., Yamada, S., Kimura, R., Yajima, T., and Kashimoto, K. Long-acting analogue of vasoactive intestinal peptide, [R15, 20, 21, L17]-VIP-GRR (IK312532), protects rat alveolar L2 cells from the cytotoxicity of cigarette smoke. *Regul Pept* (2004) 123: 193-199.

[56] Watanabe, T., Masuo, Y., Matsumoto, H., Suzuki, N., Ohtaki, T., Masuda, Y., Kitada, C., Tsuda, M., and Fujino, M. Pituitary adenylate cyclase activating polypeptide provokes cultured rat chromaffin cells to secrete adrenaline. *Biochem Biophys Res Commun* (1992) 182: 403-411.

[57] Vaudry, D., Rousselle, C., Basille, M., Falluel-Morel, A., Pamantung, T. F., Fontaine, M., Fournier, A., Vaudry, H., and Gonzalez, B. J. Pituitary adenylate cyclase-activating polypeptide protects rat cerebellar granule neurons against ethanol-induced apoptotic cell death. *Proc Natl Acad Sci U S A* (2002) 99: 6398-6403.

[58] Dejda, A., Sokolowska, P., and Nowak, J. Z. Neuroprotective potential of three neuropeptides PACAP, VIP and PHI. *Pharmacol Rep* (2005) 57: 307-320.

[59] Liu, G. J., and Madsen, B. W. PACAP38 modulates activity of NMDA receptors in cultured chick cortical neurons. *J Neurophysiol* (1997) 78: 2231-2234.

[60] Said, S. I., Dickman, K., Dey, R. D., Bandyopadhyay, A., De Stefanis, P., Raza, S., Pakbaz, H., and Berisha, H. I. Glutamate toxicity in the lung and neuronal cells: prevention or attenuation by VIP and PACAP. *Ann N Y Acad Sci* (1998) 865: 226-237.

[61] Brenneman, D. E., Hauser, J. M., Spong, C., and Phillips, T. M. Chemokine release is associated with the protective action of PACAP-38 against HIV envelope protein neurotoxicity. *Neuropeptides* (2002) 36: 271-280.

[62] Fairbairn, D. W., Carnahan, K. G., Thwaits, R. N., Grigsby, R. V., Holyoak, G. R., and O'Neill, K. L. Detection of apoptosis induced DNA cleavage in scrapie-infected sheep brain. *FEMS Microbiol Lett* (1994) 115: 341-346.

[63] Joseph, S. K., and Ryan, S. V. Phosphorylation of the inositol trisphosphate receptor in isolated rat hepatocytes. *J Biol Chem* (1993) 268: 23059-23065.

[64] Eikvar, L., Tasken, K. A., Eskild, W., and Hansson, V. Protein kinase C activation and positive and negative agonist regulation of 3',5'-cyclic adenosine monophosphate levels in cultured rat Sertoli cells. *Acta Endocrinol (Copenh)* (1993) 128: 568-572.

[65] Favata, M. F., Horiuchi, K. Y., Manos, E. J., Daulerio, A. J., Stradley, D. A., Feeser, W. S., Van Dyk, D. E., Pitts, W. J., Earl, R. A., Hobbs, F., Copeland, R. A., Magolda,

R. L., Scherle, P. A., and Trzaskos, J. M. Identification of a novel inhibitor of mitogen-activated protein kinase kinase. *J Biol Chem* (1998) 273: 18623-18632.

[66] Eichholtz, T., de Bont, D. B., de Widt, J., Liskamp, R. M., and Ploegh, H. L. A myristoylated pseudosubstrate peptide, a novel protein kinase C inhibitor. *J Biol Chem* (1993) 268: 1982-1986.

[67] Douglas, W. H. J., and Kaighn, M. E. Clonal isolation of differentiated rat lung cells. *In Vitro* (1974) 10: 230-237.

[68] Douglas, W. H. J., and Farrell, P. M. Isolation of cells that retain differentiated functions in vitro: properties of clonally isolated type II alveolar pneumocytes. *Environ Health Perspect* (1976) 16: 83-88.

[69] Onoue, S., Ohmori, Y., Matsumoto, A., Yamada, S., Kimura, R., Yajima, T., and Kashimoto, K. Structure-activity relationship of synthetic truncated analogues of vasoactive intestinal peptide (VIP): an enhancement in the activity by a substitution with arginine. *Life Sci* (2004) 74: 1465-1477.

In: Neuronal Cell Apoptosis Research
Editor: Relif J. Shandling, pp. 75-96
ISBN 1-60021-456-8


*Chapter 3*

# MECHANISMS INVOLVED IN E2F1 INDUCED NEURONAL APOPTOSIS

***Mercè Pallàs*[1,*]*, Anna M. Canudas*[1]*, Ester Verdaguer*[2]*,***
***Jaume Folch*[3]* and Antoni Camins*[1]

[1]Unitat de Farmacologia i Farmacognòsia, Facultat de Farmàcia, Universitat de Barcelona, Nucli Universitari de Pedralbes, E-08028 Barcelona, Spain;
[2]Departament de Farmacologia i Toxicologia, IIBB-CSIC, IDIBAPS, Rossello 161, Planta 6, 08036 Barcelona, Spain;
[3]Unitat de Bioquímica, Facultat de Medicina i Ciències de la Salut, Universitat Rovira i Virgili. C./ St. Llorenç 21, 43201 Reus (Tarragona), Spain.

## ABSTRACT

One of the challenges in the next few years will be to clarify the apoptotic routes in neurons in order to design potential drugs for the treatment of Alzheimer's disease, Parkinson's disease and other neurological disorders. Although an enormous effort has been made in the past few years and it has been evidenced that the mitochondria comprise a key component of the neuronal apoptotic route, it seems that in addition to the mitochondria other intracellular components are implicated in this process. With reference to this, it has been proposed that DNA damage and reentry into the cell cycle could constitute a common pathway in the apoptotic process in neurological diseases. This hypothesis is supported by the existing data in regard to neurodegenerative disorders in the brain of patients who show an increase in cell cycle protein expression compared with control brains. Indeed, studies performed in neuronal cell preparations suggest that reentry into the cell cycle and, concretely, increased expression of the transcription factor E2F1 may be the prominent pathway involved in the regulation of

* Correspondence concerning this article should be addressed to Dra. Pallàs, Unitat de Farmacologia i Farmacognòsia, Facultat de Farmàcia, Universitat de Barcelona, Nucli Universitari de Pedralbes, E-08028 Barcelona, Spain. E-mail: pallas@ub.edu

apoptosis. We also summarize the potential routes involved in E2F1 induced apoptosis, p53 dependent and p53 independent. In this last hypothesis, multiple pathways have been evidenced, including the activation of caspases. Finally, given the increasing experimental data on the neuroprotective and antiapopotic effects of CDKs inhibitory drugs, among them flavopiridol, it is logical to predict the potential application of these drugs for the treatment of neurological disorders.

**Keywords:** Neurodegeneration, apoptosis, cell cycle, E2F1, mitochondria, DNA damage

## INTRODUCTION

The process of neuronal cell death remains unexplained, and neuroscientists are attempting to understand the potential pathways involved in this process. Several questions remain to be answered, for instance, What are the molecular pathways involved in the process of neuronal cell death? Is it only the mitochondria involved in the apoptotic process? Or does cross-talk with other targets exist? However, apart from the potential intracellular targets, it is clear that aging is probably the main factor or cause that predisposes to neurodegenerative disease (Cash et al. 2002; Chong et al., 2005; Atwood et al., 2005). Because life expectancy is increasing, the probability that people will suffer a neurodegenerative disease such as Alzheimer's disease (AD) and Parkinson's disease (PD) will increase. Thus, one of the objectives of the pharmaceutical industry will be the development of effective drugs for the treatment of AD and PD. For this purpose, it is of vital importance to understand the process of neuronal cell death and, more specifically, to understand the mechanisms of apoptosis (Shimohama, 2000; Reed, 2001; Gastard et al. 2003; Friedlander, 2003; Lockshien and Zakari, 2004). As mentioned above, there are many important questions that still remain unanswered, but nowadays neuroscientists are progressively gaining more information about the process of neuronal loss.

In this review, we describe one potential pathway of the probably multiple mechanisms which are involved in apoptotic neuronal death, namely reentry into the cell cycle. It is well-known that neurons are quiescent cells which are not able to proliferate. However, at the beginning of 1990, a new pathway was suggested whereby neurons could reenter or try to reenter the cell cycle (Nagy et al. 1998; Manccioni et al. 2001; Reed, 2001; Nishioka and Arnold, 2004; Frappart et al. 2005; Neve et al. 2006). Thus, the neuronal oncogenic hypothesis was proposed in order to explain neuronal demise, mainly in AD. Furthermore, we also discuss the hypothesis, suggested in the last few years, on the link between reentry into the cell cycle and apoptosis.

## APOPTOSIS IN NEURONS: A MITOCHONDRIAL PERSPECTIVE

In an attempt to simplify the process of neuronal demise, a classification of the mechanisms of neuronal cell death into three types has been proposed (Kerr et al., 2002; Lockshien and Zakari, 2004). Type I is the classical apoptosis (programmed cell death)

implicated in physiological processes, for instance in the development of the central nervous system, where some neurons die by apoptosis. The main cystein proteases activated in this process are caspases (Reed, 2001). Type II has been classified as autophagic cell death based on and characterized by the formation of vacuoles and the activation of cathepsins. Finally, type III neuronal cell death is the necrotic process, which is known as an uncontrolled mechanism of cell death when the rapid drop in mitochondrial ATP levels leads to breakdown of cell membranes and death (Nicotera and Leist, 1997). Because neurodegenerative disease has its own clinical course but usually progresses slowly for more than 10 years, it is logical to think that apoptosis will be the main process implicated in neuronal demise. In this regard, it is widely accepted that the mitochondria orchestrate the neuronal apoptotic process (Figure 1). Although the apoptotic route is complex, Kroemer and colleagues were the first to propose a clear apoptotic mechanism gated to mitochondria (Kroemer et al. 1998). They proposed a classification of the process into three phases. The first step is known as the inductor phase, where for instance ROS generation, intracellular calcium increase and other proapoptotic stimuli converge towards the mitochondria. In a second phase, called the effector phase, changes in the mitochondrial transition pore lead to the release of pro-apoptotic proteins such as cytochrome c (Cyt c), apoptosis inducing factor (AIF), endonuclease G and the second mitochondria-derived activator of caspase Smac/DIABLO (Van Loo et al. 2002). The first protein identified with a pro-apoptotic function was Cyt c; however, this protein has a physiological function normally involved in electron transport. Cyt c is released from the mitochondria to the cytoplasm through the permeability transition pore (PTP). Many reports suggest that release of Cyt c is mediated by a prior loss of mitochondrial transmembrane potential; however, numerous studies demonstrate that Cyt c release is possible without changes in the mitochondrial transmembrane potential. Once Cyt c is released to the cytoplasm, it triggers the formation of the apoptosome, a complex composed of Cyt c, the apoptotic protease activating factor-1 (Apaf-1) and dATP. This complex helps to activate the initiator procaspase-9, and in successive steps the caspase cascade is amplified by activation of executioner caspases such as caspase-3, caspase-6 and caspase-7 (Garcia-Calvo et al. 1998; Troy and Salvessen, 2002). This latter phase is known as the degradation phase, during which the nucleus is fragmented. Therefore, all these mitochondrial proteins that are released by mitochondria are implicated in the regulation of downstream components of the cell death machinery. It is important to remark that the most widely studied and best characterized caspase in neurons is caspase-3, for which a key role in DNA fragmentation has been demonstrated.

However, the mitochondrial pathway is really a more complex route, because the PTP aperture is regulated by the Bcl-2 family of proteins which also help release pro-apoptotic factors from the mitochondrial intermembrane space into the cytoplasm. Furthermore, members of the Bcl-2 family are associated with the outer mitochondrial membrane, and proapoptotic members of this family could oligomerize and induce the formation of mitochondrial pores, leading to the release of mitochondrial proteins (Van Loo et al. 2002).

The Bcl-2 family is classified into three subfamilies. Firstly, there is a subfamily of anti-apoptotic proteins, including among other Bcl-2, Bcl-$x_L$ and Bcl-w. Another subfamily contains a group of pro-apoptotic proteins including Bax, Bak and others that induced the release of Cyt c. Finally, a subfamily exists that includes other pro-apoptotic proteins like Bad, Bid, and Bim (Van Loo et al. 2002).

Although it is well accepted that the activation of what has been called the 'intrinsic apoptotic pathway' could contribute to the process of neuronal demise, the picture is more complicated by the evidence that caspase inhibitors only attenuate or delay the process of neuronal cell death but do not prevent it completely. This suggests that additional, mitochondria-independent routes involved in neuronal cell death could exist. In addition, the mitochondria may release, through the PTP, other pro-apoptotic proteins that by themselves regulate a caspase-independent apoptotic pathway (Stefanis et al. 1999).

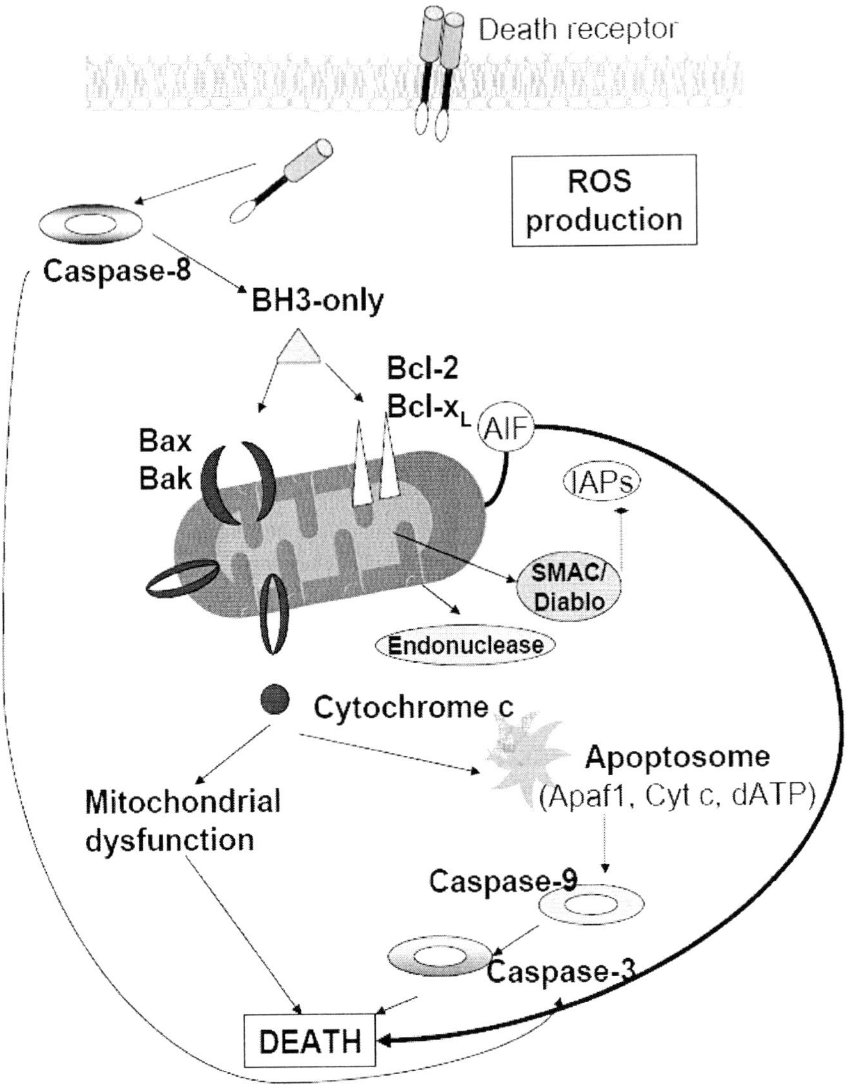


Figure 1. Plausible model of how mitochondria orchestrate the neuronal apoptotic process. Extrinsic apoptotic pathway was mediated by death receptor/caspase-8 activation. Meanwhile intrinsic pathway goes through cytochrome C release and caspase-3 activation (See text).

## Apoptosis in Neurons: The Role of DNA Damage

Although the mechanism of neuronal cell death in neurological diseases is still obscure, as noted above, it has been hypothesized that cell death is not only mediated by mitochondrial perturbation and the release of proapoptotic proteins. Furthermore, the cell cycle becomes a key component in the machinery of neuronal loss (Smith et al., 2004). This hypothesis has been called the two hit hypothesis of AD and suggests that in this neurodegenerative disease neurons suffer two stimuli: oxidative stress, and the reentry into the cell cycle (Zhu et al. 2004; Weber et al. 2005). This hypothesis is supported by several evidences: first, it is widely known that the brain is highly vulnerable to reactive oxygen species, probably due to the antioxidant enzymes superoxide dismutase, catalase, and others displaying a low activity in AD and also in ageing (Chong et al. 2005). Furthermore, an increase in reactive oxygen species constitutes a common event in all neurodegenerative diseases. The second hit hypothesis, the reentry into the cell cycle, is based on the evidence that in the brain of patients with AD, PD, and other neurological diseases the presence of an increase in cell cycle markers has been demonstrated (Arendt et al. 1998; Busser et al. 1998; Bowser et al. 2002; Herrup et al. 2002; Jordan-Sciutto et al. 1999; Jordan-Sciutto et al. 2001; Jordan-Sciutto et al. 2003). At first, the increase in the expression of cell-cycle markers in the brain of AD patients was a surprise, since neurons do not proliferate and they are in a quiescent state (also called cell cycle G0 phase). However, the over-expression of a wide range of cell-cycle proteins, namely cyclin-dependent kinases (CDKs), has been demonstrated, and these are associated with cyclins in neurons of AD patients but not in age-matched controls (Arendt et al. 1998; Jordan-Sciutto et al. 1999). CDKs constitute a family of serine/threonine kinases which regulate the cell division. As discussed earlier, neurons are in a G0 phase of the cell cycle; however, the cell cycle is divided into four phases: G1 (first gap), S (DNA synthesis), G2 (second gap), and M (mitosis) and CDKs regulate the progression through these phases by a highly complex process (Figure 2, 3). Thus, Cdk4, 6 associated with cyclin D1, D2, D3 regulated the transition of Go/G1 phase. Next, in the G1 phase Cyclin E is induced and associates with cdk2. Cyclin E jointly with cdk2 is necessary for the transition into S phase. At the end of G1 phase there is the restriction point the retinoblastoma protein pRb. Cyclin D/cdk4-6 complex is necessary for the cell to overcome the restriction point pRb (Yu et al. 2005). This complex is responsible for pRb phosphorylation. In normal conditions pRb is hypo-phosphorylate and it binds to E2F/DP transcription factors. When the cell cycle has to start S phase, CyclinD/cdk4-6 complex phosphorylates pRb in order to dissociate E2F/DP from it. When E2F1 is released activates several genes which are necessary for the transcription and protein synthesis during S phase. Therefore, the cells which are able to overcome the restriction point pRb will continue their way through the cell cycle. On the other hand, cells which are not able to enter in S phase will be kept in G0 phase or the quiescent phase. The synthesis of DNA is associated with the activation of cyclin A/cdk2 complex. When DNA replication is completed, cyclin A forms a complex with cdk1 and drives the cell through G2 phase. Finally in the G2/M phase, cdk1 form a complex with cyclin B and the cell enter in mitosis.

However, the question that remains to be determined is: What are the molecular pathways responsible for the induction of cell-cycle proteins? As we suggested above, probably the first event responsible for reentry into the cell cycle is oxidative stress (Klein et al. 2002, 2003; Smith et al. 2004). This hypothesis is evidenced by the study performed by Klein and colleagues where, evaluating the process of neuronal cell death in a Harlequin mouse model, it was found that oxidative stress in implicated in neuronal cell cycle reentry. Harlequin is an X-linked mutation and the mice developed progressive ataxia associated with the death of both granule and Purkinje cells in the cerebellum, as well as photoreceptor cells of the retina (Klein et al., 2002). The authors demonstrated changes in oxidation damage repair and also alterations of cell cycle control. Therefore this model permits us to hypothesize that cell cycle reentry is strictly associated with cell death in this mouse model, as is indicated by the finding that all apoptotic neurons express S-phase markers. Thus, oxidative stress associated with DNA damage could constitute a critical event in the process of reentry into the cell cycle. Furthermore, this hypothesis is supported by several evidences, for instance, the brains of patients with AD show a decrease or attenuation in neuronal DNA repair. The brain neurons of these patients also show a decrease in the expression of Mre11 complex which is involved in DNA damage repair, DNA replication and cell cycle checkpoints (Jacobsen et al. 2004). So, Jacobsen and colleagues suggest that Mre11 complex is lost in this neurodegenerative disease. Inability to activate a functional Mre11 complex may contribute to neuronal loss, especially during an attempt to reenter the cell cycle from a post-mitotic state. Kruman and colleagues evidenced the critical role of DNA damage on the process of reentry into the cell cycle. In this study, neurotoxins which do not induce DNA damage such as staurosporine and colchicine did not initiate the expression of cell-cycle proteins. Likewise, they demonstrated a key role of ataxia telangiectasia mutated (ATM), a cell cycle checkpoint. Pharmacological inhibitors of ATM such as caffeine and wortmanin show neuroprotective properties against neurotoxins, which induces DNA damage in neurons and prevents both apoptosis and cell-cycle reentry triggered by DNA (Kruman et al, 2004). Another question that remains is how, after the DNA damage, neurons start to synthesize DNA and trigger the apoptotic process. To answer this question Copani and colleagues investigated the role of several polymerases which are involved in the biochemistry of eukaryotic DNA replication. Three different DNA polymerases (pol), polα, polδ, and polε are implicated in the process of proliferation in mammalian cells. Polα, initiates de novo DNA replication by synthesizing short RNA-DNA primers. Polδ and pol ε can function as DNA repair enzymes, correcting the potential mismatch errors during replication (Copani et al., 2002). Therefore, the coordinated regulation of the activity of multiple DNA polymerases is fundamental in the process of DNA replication in dividing cells. Notably, polβ, participates in base excision repair, a pathway which replaces oxidized bases. The precise repertoire of DNA polymerases that are expressed by neurons is unknown. Therefore, in neuronal DNA synthesis, this polymerase family might be a potential source of replication errors that are able to trigger the activation of a p53-dependent death pathway (Figure2) (Copani et al. 2001). So DNA polymerases could be another potential death signal implicated in the neuronal apoptotic route.

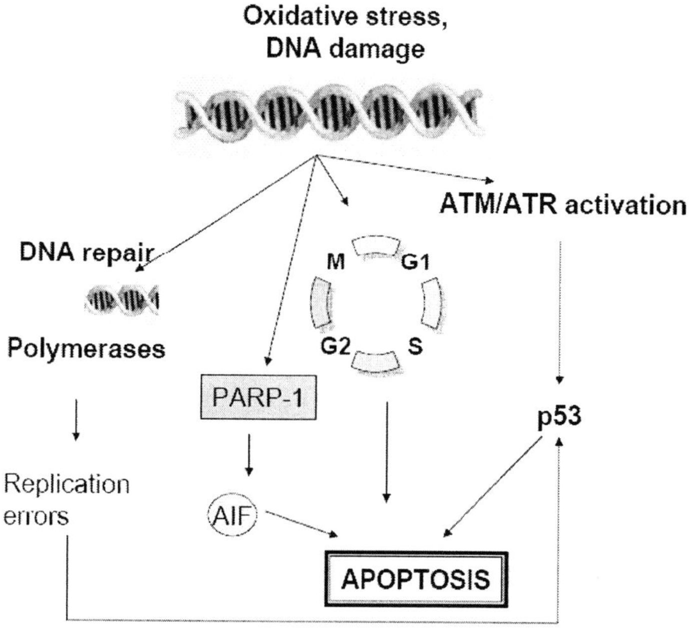


Figure 2. Apoptotic process following DNA damage. The cell cycle as a key component of neuronal loss, jointly with ataxia telengectasia mutated kinase (ATM), p53, AIF and PARP-1 activation.

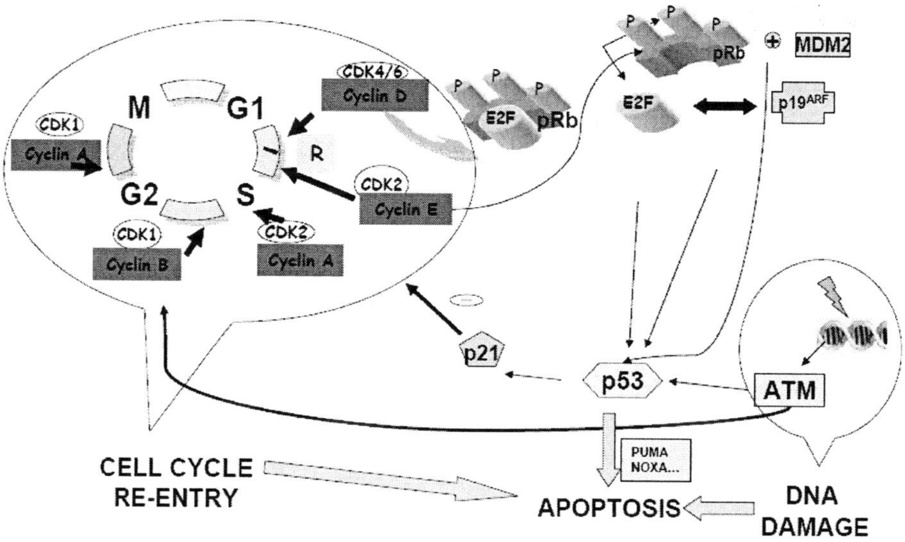


Figure 3. Neuronal reentry into the cell cycle in neurons as a potential step in neuronal demise: CDKs activation orchestrates the death process through the phosphorylation of the retinoblatoma protein (pRb), release of E2F1 and activation of p53 death signaling.

Poly (ADP-ribose) polymerase-1 (PARP-1) is a nuclear enzyme that is immediately activated by DNA damage. The activation of PARP-1 following DNA damage is another important target in neuroprotection because there is an amount of data to suggest that PARP-1 modulates the process of neuronal apoptosis (Dawson and Dawson, 2004). Wang and colleagues demonstrated that neuronal damage following the activation of the *N*-methyl-D-aspartate (NMDA) receptor is accompanied by the formation of peroxynitrite ion a powerful oxidant (Wang et al. 2004). Thus, Dawson and colleagues demonstrated a mechanism implicated in glutamate excitotoxicity through the formation of NO and free radicals via PARP-1. So, NMDA stimulation can damage DNA and activate PARP-1. Furthermore, PARP-1 activation by NMDA evokes a mitochondrial release and nuclear translocation of AIF, a caspase independent mediator of neuronal cell death. Therefore, Dawson and colleagues suggest that AIF following PARP-1 activation substitute the activation of caspases and PARP-1 constitute an upstream signal involved in mitochondrial regulation of apoptotic process following DNA damage (Figure 2).

## Apoptosis in Neurons: the Role of Cell Cycle

The next step is to comment on how mitochondrial alteration and DNA damage could constitute two intracellular targets implicated in the regulation of apoptosis (Figure 3). One way could be via the expression of CDKs (Lim and Qi, 2003). Although the mechanism implicated in the way CDKs activation orchestrates the death process is not completely known, one potential pathway that has been suggested may be through the phosphorylation of the retinoblatoma protein (pRb) (Atwood et al. 2005). In what follows, we will comment on the importance of neuronal reentry into the cell cycle in neurons as a potential step in neuronal demise.

It is also well established that several human neurodegenerative pathologies such as AD, PD, Amyotrofic Lateral Sclerosis are accompanied by an increase in the expression of cell-cycle proteins (Jordan-Sciutto et al. 2002; Andorfer et al. 2005; Appet-Collin et al. 2006). For instance, Jordan-Sciutto et al. observed the presence of proteins involved in cell-cycle activation in the brains of patients with PD, as well as changes in the expression of the pRb (Jordan-Sciutto et al. 2002). So, it is possible to suggest that reentry into the cell cycle might be a common component of the process of neuronal cell death. In this way, neurons will be able to change from a quiescent state to G1 phase leading to aberrant proliferation without to complete a neuronal duplication. Yang and colleagues, using the technique of fluorescent in situ hybridization (FISH), have demonstrated in neurons of AD patients a cell-cycle progression, because this methodology makes it possible to distinguish DNA replication from other synthetic events such as DNA repair (Yang et al. 2001). Furthermore, the same authors have demonstrated a substantial expression of cell cycle markers at the beginning of this neurodegenerative disorder in the brains of patients with AD showing a mild cognitive impairment (Yang et al. 2003). These results are important because they indicate that the cell cycle could be an early and crucial step in the AD pathology. Furthermore, they found an increase of cyclin B in neurons of AD patients and hypothesized that such neurons nearly complete DNA replication and that the process of neuronal division is probably stopped before the M phase. Indeed, in AD brains, Jordan-Sciutto and colleagues observed an increase in pRb and E2F immunoreactivity, with pRb predominately located in the nucleus and E2F1 in the cytoplasm. In patients with ALS, Ranganathan and Bowser evidenced the expression of cell-cycle proteins in the spinal cord, motor and sensory cortex. Those authors evidenced a nuclear accumulation of hyperphosphorylated pRb with a concurrent increase in cytoplasmic E2F1 immunoreactivity, suggesting a role for aberrant activation of G1 to S phase regulators without progression to DNA synthesis. In motor neurons, they found an increase in the levels of cyclin D and cdk4. In the brain of patients with PD, Jordan-Sciutto demonstrated in the substantia nigra an increase in pRb phosphorylation which was co-localized with Lewy bodies (Jordan-Sciutto et al. 2001, 2003). These data give support to the implication of cell cycle in the neurodegenerative process of PD. In another study, Love demonstrated the neuronal expression of cdk2, cdk4 and cyclins in the brain of patients with focal brain infarction. These data suggest that after an ischemic process some neurons in the brain reenter the cell cycle. Therefore, cell cycle reentry in human neurons will be important in different neurological disorders.

On the other hand, studies performed in neuronal cell preparations demonstrated that after the addition of β-amyloid (an experimental model of AD) an increase in the expression of cell-cycle proteins and DNA synthesis was found (Copani et al. 2002; Giovanni et al. 1999, 2000; Jordan-Sciutto et al. 2001). Furthermore, blockade of the G1/S transition using a cyclin D1 antisense or a dominant-negative mutant of cdk2 prevented both β amyloid-induced DNA replication and apoptosis, so these data in neuronal cell preparations give support to the importance of the cell cycle in the regulation of neuronal apoptotic process (Copani et al. 2001, 2002). Since, it is improbable that neurons completely duplicate, this process of reentry into the cell cycle has been called as an abortive attempt at cell division. However, analysis by fluorescent-activated cell-sorting techniques has shown that neurons could synthesize new DNA (Copani et al., 1999, Kruman et al 2004, Verdaguer et al.2002, 2004, 2005). Therefore, these studies performed in neuronal cell cultures indicate that differentiated neurons can increase the expression of cell-cycle proteins (e.g., cyclins) and try to synthesize DNA, which leads to death by apoptosis and not to proliferation (Giovanni et al., 1999, Verdaguer et al. 2002, 2004; Otsuka et al, 2004; Wu et al. 2004; Wen et al. 2005). The main question that remains to be answered is whether reactivation of the cell cycle is an obligatory step in the apoptotic pathway or not. In other words, is reactivation the cause or the consequence of neuronal cell death? Since pharmacological inhibitors of cell cycle such as flavopiridol, roscovitine and oloumicine protect from this neurotoxic stimuli, these data confirm the importance of this pathway in neuronal cell death (Verdaguer et al. 2004, 2005; Wu et al. 2004). Furthermore, Appert-Collin and colleagues demonstrated that CDKs inhibitors suppress the death of motoneurons after neurotrophic deprivation. Park and colleagues, using PC12 cells, also demonstrated that G1/S cell cycle blockers such as mimosine and deferoxamine and cyclin-dependent kinases inhibitors attenuate the death of nerve growth factor (NGF)-deprived PC12 cells and sympathetic neurons (Park et al. 1997; 1998). The neuroprotective effects of CDKs inhibitors have been demonstrated in other insults (Verdaguer et al. 2002, 2004). For instance, after the stimulation of glutamate receptors with kainic acid or glutamate, flavopiridol attenuates the process of apoptosis (Padmanabhan et al. 1999; Martin-Romero et al. 2000; O'hare et al. 2000; Verdaguer et al. 2004; Wu et al. 2004). Besides, the administration of kainic acid in rats induces the increase of cell-cycle proteins (Park et al. 2000). Indeed, the administration of flavopiridol in rats attenuates the neuronal damage mediated by traumatic brain injury and inhibits the glial proliferation and attenuation of microglial activation (Cernak et al. 2005). In cerebellar granule cells, CDKs inhibitors attenuate the apoptotic process mediated by serum potassium deprivation and the neurotoxin colchicine (Jordà et al. 2003; Verdaguer et al. 2005). However, colchicine-induced apoptosis is mainly mediated by cdk5, a CDKs which is not implicated in cell-cycle regulation, and this neurotoxin does not cause DNA damage (Kruman et al. 2004). Flavopiridol is also able to protect from an ischemic process in neuronal cell cultures and after the administration in rats. In fact flavopiridol protects cortical neurons from hypoxia-induced apoptosis and decreases the expression of cyclin D1 (Wu et al. 2004). Wang and coworkers propose that one of the intracellular signals implicated in the cell-death process in ischemia is the activation of CDKs (Wang et al. 2004; Wen et al. 2005). After a focal or global ischemic insult they demonstrated an increase in the expression of cell-cycle regulators such as cyclin D1 and Cdk4. Besides, the administration of flavopiridol, significantly protects CA1 neurons from delayed neuronal death. Osuga and coworkers also demonstrated that the intracerebroventricular administration of flavopiridol at the concentrations of 100 or 500 μM to the animals furthermore exerts behavioral improvements and long-term neuroprotection. Indeed, the compound 219476, which is a highly Cdk4/Cyclin D1 inhibitor complex, shows neuroprotective properties in a model of ischemic damage (Wen et al. 2005). Therefore, this study performed by Wen and coworkers corroborates that neuronal apoptosis induced by transient focal cerebral ischemia involves the induction of cell-cycle related proteins and, potentially, new DNA synthesis. However, only a few researchers have demonstrated an unequivocal evidence of DNA synthesis by degenerating neurons in vivo (Yang et al., 2001, Klein et al., 2002). For example, Katchanov and coworkers, in a stroke model they found a high expression of cdk4, cyclin D1, cyclin B1 and E2F1. Besides, they demonstrated that only 1% of the TUNEL-positive nuclei after ischemia incorporated BrdU. The small percentage of BrdU- and TUNEL-double-positive cells suggests that neurons very rarely pass the G1/S-phase checkpoint, even reenter in the cell cycle. Probably, this hypothesis could be applied to all stimuli.

The fungal toxin, 3-nitroproprionic acid (3-NPA) is an irreversible inhibitor of succinate dehydrogenase, a component of the electron transport chain, and leads to loss of ATP synthesis. Previous studies demonstrated a neuronal loss in cultures after a treatment with 3-NPA. The addition of flavopiridol jointly with 3-NPA to neuronal cell cultures offers a partial neuroprotection (Park et al. 2000). However, addition of flavopiridol jointly with MK-801 provides higher neuprotection. A dose response analysis of flavopiridol indicates that a concentration of 1–3 μM provides maximum protection when coadministered with MK-801. These results suggest that neuronal death mediated by excitotoxic stress may include cell cycle regulators at both early and late times. Once the neurons are protected from early excitotoxic death by inhibitors such as MK-801 or CDKs inhibitors may confer protection from delayed death.

Zhang and coworkers working with a mouse model of Niemann-Pick disease demonstrated that the intracerebroventricular administration of CDKs inhibitors attenuated significantly the hyperphosphorylation of tau, and ameliorated motor defects in this mouse model of the disease. Therefore, these data suggest that CDKs activities are involved in the mouse neuropathology of this disease model (Zhang et al. 2004).

## E2F1: THE POTENTIAL LINK BETWEEN REENTRY INTO THE CELL AND APOPTOSIS

After the evidence that activation of cell-cycle increase in the expression of cdk4/cdk6 and cyclin D in neurons in vivo and in vitro, another question remains to be answered: What is the link between the cell cycle and apoptosis? Above we suggest that neuronal division does not continue after neurotoxic stimuli. Thus, although neurons may start to synthesize DNA, the whole process of DNA synthesis cannot be completed. The reason why neurons are not able to finish this DNA synthesis is unknown, but the most plausible explanation is that several factors are involved in cell-cycle progressions such as E2F1, which could be the responsible element for or could orchestrate the apoptotic process of neuronal cell death (Motonaga et al. 2001; Smith et al. 2003; Ueberham and Arent, 2005).

E2F1 belongs to the family of transcription factors that coordinate the expression of genes during the cell-cycle progression. Moreover, it has been demonstrated that some factors like E2F1 could interfere in programmed cellular death. Eight members of E2F family members have been identified (E2F1-E2F8) (Dimova et al. 2005; Stanell et al. 2006). The physiological activity of E2F1-5 is regulated by the binding of retinoblastoma protein Rb and the related p107 and p130 proteins (the so-called pocket protein family). The function of these families of proteins is to arrest the cell cycle in the G1 phase. The retinoblastoma protein is at present the best characterized regulator of E2F family members. It is well known that the E2F transcription factor family leads to an increase in the genes involved in the transition phase $G_1$-S and is besides also implicated in the regulation of genes involved in apoptosis. Its structure is composed by two subunits: one is encoded with genes of the E2F family and another by genes and of the DP family (Cooper –Kuhn, 2002; Stanelle et al. 2006).

The E2F family is classified into:

- E2F1, E2F2, E2F3: these factors are associated with pRb and they are powerful trascriptional activators expressed during the cell cycle.
- E2F4 and E2F5: is associated with pRb and with pocket proteins function as repressors of recruiting pocket proteins.
- E2F6 and E2F7: these factors do not interact with pocket proteins but work like a negative controller of the transcription process. However, the majority of functions of the new E2F members are currently under investigation.

In reference to the pathways involved in apoptosis, several pathways have been suggested, such as:

p53 dependent, mitochondrial release of proteins, activation of proapoptotic genes, Apaf-1/bax and other signals for example NFκβ (Strachan et al. 2005).

E2F1 is a key modulator of apoptosis demonstrated by the evidence that neurons lacking this protein are protected from neurotoxin stimuli, such as serum potassium deprivation in cerebellar granule cells or stimulation of AMPA/kainate receptors using the neurotoxin kainic acid and ischemia (Verdaguer et al. 2004; MacManus et al. 1999; MacManus et al. 2003). Besides, in cortical cultures and hippocampal slices, oxygen-glucose deprivation induces apoptosis is attenuated in cultures lacking E2F1 (Hou et al. 2000, 2001; Gendron et al. 2001). Neuronal cultures deficient in E2F1 are transiently protected from death evoked by β-amyloid. In addition, these results are in accordance with observations that E2F1 expression itself is sufficient to induce death of cortical neurons in the absence of any apoptotic stimuli and that expression of dominant negative DP1 protects neurons from β-amyloid (Giovanni et al. 2000). Taken together, along with findings that Cdk4/6 activity is required for death of β-amyloid treated cortical neurons, these results suggest a model by which Cdk4/6 activation, the consequent phosphorylation/inactivation of pRb, and activation of E2F1-containing complexes contribute to the death signal evoked by Aβ treatment (Giovanni et al. 1999).

As mentioned above, one potential apoptotic route regulated by E2F1 is through the induction of p53 (Figure 3). Although the pathways involved in p53 activation are extremely complex, a role of DNA damage and cell cycle has been proposed. Thus, it has been suggested that Rb protein can be complexed with MDM-2 and p53 and lead a high p53 activity and apoptosis (Xie et al., 2006). Furthermore, high levels of active E2F1 not bound to Rb switches the p53 response from G-1 arrest to apoptosis. Once p53 is activated, it stimulates the synthesis of the p21 protein, which inhibits cyclin E-cdk2 activity, and this in turn acts upon the Rb-MDM-2 complex that promotes p53 activity and apoptosis. This enhances p53 activity in the same way that the p53-MDM-2-Rb complex increases p53 function and is proapoptotic. Other alternative pathways are mediated through the induction of ADP-rybosylation factor (ARF), E2F1 transactivate directly ARF promoting p53 stability and thus inducing apoptosis (Philips et al. 1999). E2F1 upregulates p19 ARF which then stabilizes p53 levels, possibly through a direct or indirect function of MDM2, for example, by modulating its ubiquitin ligase activity, thereby increasing the levels of the p53 protein. In neuronal preparations it has been suggested a potential route of p53-induced apoptosis is through ATM activation after DNA damage (Mcmurray 2005). ATM regulates p53 stabilization and neuronal death. Thus, ATM function is required for neuronal apoptosis, and a complex relationship between ATM, E2F1 and p53 in response to DNA damage exists (Kruman et al. 2004). However, several data indicate that p53 is neither required for the death of cortical neurons evoked by β-amyloid nor for E2F1-induced apoptosis, through cell cycle entry, in CGNs after potassium withdrawal (O'hare et al. 2000; Giovanni et al. 2000). So the requirement of p53 may differ through different death stimuli and E2F1 involvement, and it is currently unknown whether additional p53 family members serve to regulate this death paradigm (Cregan et al. 2004).

With reference to the apoptotic route modulated by E2F1 and p53, it has been suggested that E2F1 may be localized upstream of p53 leading to p53 stabilization and upregulation or downstream (DNA damage induces p53 and re entry into the cell cycle). Then p53 would activate apoptosis by its strong pro-apoptotic function exerted through activation of PUMA or NOXA (Furukawa et al. 2001; Fortin et al. 2004). However, it seems that the role of NOXA in neuronal apoptosis may be minor. On the contrary, Cregan and coworkers demonstrated in neuronal cell cultures that expression of PUMA (p53 upregulated modulator of apoptosis) is strongly correlated with apoptosis, and ectopic expression of PUMA itself was sufficient to induce apoptosis. Furthermore, PUMA-deficient neurons were substantially more resistant to cell death induced by p53 overexpression. All these data demonstrate that PUMA is a potent inducer of neuronal apoptosis mediated by p53. Hershko and Ginsberg demonstrated in NIH3T3 cells that PUMA, apart from p53, is also regulated by E2F1 (Hershko and Ginsberg, 2003). However the role of E2F1 in neuronal regulation of PUMA remains to be demonstrated.

E2F1 could also regulate the apoptotic process by a mechanism independent of p53. These processes orchestrated by E2F1 are initiated by an upregulating of the proapoptotic proteins p73 (Irwin et al., 2000). With reference to p73, it is a p53 homolog, and it has been demonstrated that E2F1 activates its transcription leading to the activation of target genes and subsequent apoptosis. The role of p73 in apoptosis is now under extensive research.

Another target of E2F1 is the induction of Cdc2 expression after potassium deprivation in cerebellar granule neurons. These data constitute a novel link in neurons between CDKs-Rb-E2F1 pathway and the mitotic kinase Cdc2 culminating in the activation of the cell-death machinery via phosphorylation of the BH3-only protein BAD (Konishi and Bonni, 2003; Becker and Bonni, 2004). Therefore, Cdc2 represents a specific target of E2F1 in postmitotic neurons.

Wang and coworkers demonstrated the regulation of E2F1 by 14-3-3γ in the apoptotic process during DNA damage. In response to DNA damage mediated by adriamycin a selective accumulation of E2F1 protein was found inducing apoptosis. By using the selective siRNA of E2F1 apoptosis was inhibited. These results are in line with the role of E2F1 in DNA damage-induced apoptosis. Furthermore 14-3-3τ mediated the apoptotic process of E2F1 by interfering with the function of an E3 ligase and inhibiting the ubiquitination and degradation of E2F1 during DNA damage (Wang et al. 2004).

A link between caspase activation and E2F1 has been suggested by Nahle and coworkers. They suggest that E2F1 simultaneously coordinates caspase expression with apoptosis and cell-cycle progression by a direct transcriptional mechanism. In another interesting study, Furukawa and coworkers, using different cell lines, demonstrated that E2F1 was able to activate caspase-9 and to initiate the caspase route without the activation of mitochondria. They proposed a new route where E2F1 activates caspase-9 through a direct increase in intracellular Apaf-1 levels (Nahle et al. 2002). Another pathway related by the inhibition of NF-kB is by the accumulation of ROS. It thereby enhances apoptosis by the suppression of superoxid dismutase (MnSOD) (Philips et al. 1999). These data are in agreement with studies performed by Hou and coworkers, who worked on neuronal cultures, and suggests that one potential target gene inhibited by E2F1 activation is the nuclear transcription factor NF-κB, a transcription factor that regulates cell survival in response to death-promoting stimuli by transcriptional activation of various apoptotic genes (Hou et al. 2001). Thus, E2F1 would exacerbate apoptosis by inhibiting cell survival signals. Therefore, 6-OHDA induced caspase-3-mediated apoptosis of cortical neurons, which was attenuated by the addition of antioxidants and by the increase in E2F1 expression (Hou et al. 2001). These findings suggest that increased expression of the transcription factor E2F1 may serve as a death signal during 6OHDA-evoked neuronal apoptosis. Furthermore, E2F1 triggers apoptosis by disrupting NF-kB signaling (Phillips et al., 1999; Hou et al, 2001). Although the literature regarding the role of NF-κB in oxidative stress-evoked neuronal apoptosis is controversial, whether pro- or anti-apoptotic, it does appear that the NF-κB function in the central nervous system is dependent upon the cell type and environment. For example, Koulich et al. (2001) showed that NF-κB was involved in the survival of CGNs, whereas Luo et al. (1999) showed that NF-κB DNA-binding activity was stimulated in response to 6OHDA in the striatal neurons, thereby implicating NF-κB in cell death.

Studies carried out on neurons have demonstrated the role of the transcription factors B- and C-myb as downstream mediators of neuronal cell death evoked by NGF deprivation and DNA damage (Liu et al. 2003). This hypothesis is supported by the fact that antisense B-myb and antisense C-myb each provided substantial protection against death evoked by NGF deprivation in PC12 cell cultures. Biswas and coworkers demonstrated that Bim, a member of the BH3 family, is a target of the classical cell cycle pathway in neuronal cells. They evidenced in PC12 that neuronal growth factor deprivation stimuli lead to cdk4 activation, consequent de-repression of E2F-regulated mybs, and induction of pro-apoptotic Bim. Notably, siRNA targeted either to B-myb or to C-myb blocked induction of endogenous Bim evoked by removal of NGF in neuronal PC12. Thus, myb is responsible for Bim induction and is a mechanism that is downstream of E2F1. Interestingly the Rb-E2F pathway associated with a derepression of the B-myb promoter activity has been associated with paraneoplastic cerebellar degeneration.

Eischen and coworkers, working with tumor cell lines, suggest that Bcl-2 is a target of E2F1. Thus, since Bcl-2 has an antiapototic effect, the suppression of this protein by E2F1 constitutes another potential factor involved in apoptosis. Furthermore, E2F1 regulates the transcription of Bcl2. Another novel death regulatory protein of E2F1 and apoptotsis has been proposed by Xie and coworkers. They demonstrated that E2F1 regulates the expression of Smac/DIABLO. This protein is released by the mitochondria and was identified as a proapoptotic factor which initiates apoptosis by binding the IAP (Inhibitor of Apoptosis Proteins), thereby relieving the inhibition on caspases. Indeed, Stanelle and coworkers reported a novel mitochondrial protein DIP (Death Inducing Protein) as an E2F1 target gene, the other mitochondrial proteins, such as Cyt c and AIF (Apoptosis Inducing Protein), also being transcriptionally regulated by E2F1. Therefore E2F1 could induce apoptosis through mitochondrial regulation.

Barbato and coworkers, working on CGNs, identified a protein called Che-1 (RNA polymerase II binding protein), which is involved in the cell cycle and furthermore interacts with Rb, regulating Rb/E2F cell cycle- promoters. Indeed, the same authors demonstrated that Che-1 interacts with neuronal cytoskeletal protein tau in CGNs. It is well known that Tau is a neuronal specific protein associated with microtubules and plays a pivotal role in the maintenance of the neuronal cytoskeleton and neurofilaments, and in the traffic of organelles. In the stimuli of serum and potassium deprivation, these authors suggest, Che-1 could be a protein implicated in the control of balance between the cell cycle and apoptosis, functions involved in apoptotic death and survival.

Thus, it seems possible to speculate that the expression of specifically the transcription factor E2F1 could constitute a common apoptotic route in neurons that could be caspase-dependent or independent.

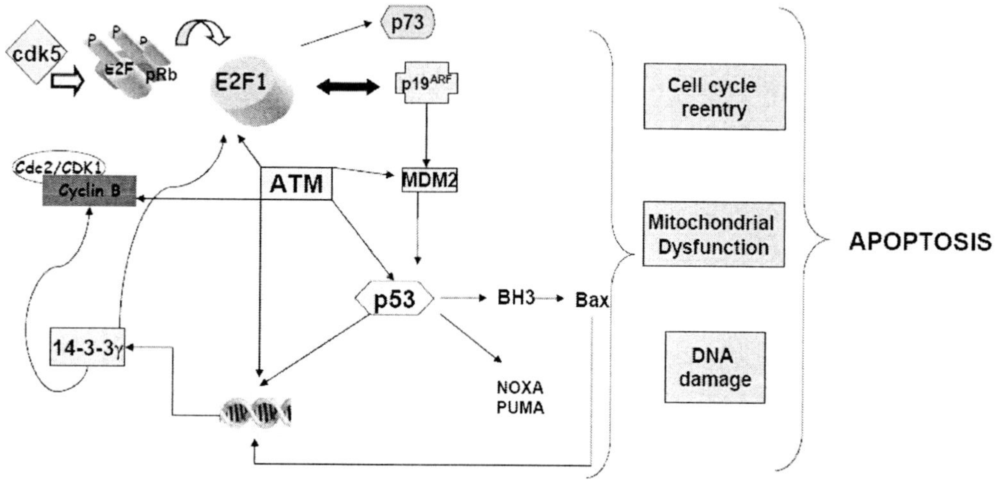


Figure 4: Hypothetic apoptotic pathways implicated in E2F1-mediated cell death: an overview.

# Concluding Remarks

In the present manuscript we summarize the data from studies performed on the role of the cell cycle and apoptosis (Figure 4). In human brains the expression of CDKs and proteins related to the cell cycle are consistent, and therefore we think that the expression of CDKs in the apoptotic route is not an artifact. From all these studies performed on human samples and in neuronal cell cultures, however, several points remain to be clarified: first, whether DNA damage is what is responsible for cell-cycle reentry through ataxia telangiectasia (ATM) or other DNA repair enzymes and, probably second, What potential pathways are involved in E2F1 mediated apoptosis? Another important point involves the recent suggestion that pRb could be phosphorylated and E2F1 released by other kinases, such as cdk5. Thus, the data we have discussed indicate a new potential apoptotic route mediated by E2F1, independently of the activation of the cell cycle. Finally, still another point currently under investigation is the potential application of CDKs inhibitors as drugs for the treatment of neurological disorders because of their capacity to inhibit E2F1 expression (Pallas et al. 2005).

# Acknowledgements

Experimental studies performed by the authors were supported by grants SAF2005-01604, SAF2005-05179 from the Ministerio de Educación y Ciencia (FEDER founds), and by FISS G03/137, FISS G03/167 and PI 041300 from the Instituto de Salud Carlos III. We thank the Generalitat de Catalunya for support to research groups 2005/SGR00893.

## REFERENCES

Andorfer, C, Acker, CM, Kress, Y, Hof, PR, Duff, K & Davies, P. (2005).Cell-cycle reentry and cell death in transgenic mice expressing nonmutant human tau isoforms. *J Neurosci, 25,* 5446-5454.

Appert-Collin, A, Hugel, B, Levy, R, Niederhoffer, N, Coupin, G, Lombard, Y, Andre, P, Poindron, P & Gies, JP. (2006). Cyclin dependent kinase inhibitors prevent apoptosis of postmitotic mouse motoneurons. *Life Sci., 79*, 484-490.

Arendt, T, Holzer, M & Gartner, U. (1998). Neuronal expression of cycline dependent kinase inhibitors of the INK4 family in Alzheimer's disease. *J Neural Transm, 105*, 949-960.

Atwood, CS, Meethal, SV, Liu, T, Wilson, AC, Gallego, M, Smith, MA & Bowen RL. (2005) Dysregulation of the hypothalamic-pituitary-gonadal axis with menopause and andropause promotes neurodegenerative senescence. *J Neuropathol Exp Neurol. 6,* 93-103.

Barbato, C, Corbi, N, Canu, N, Fanciulli, M, Serafino, A, Ciotti, M, Libri, V, Bruno, T, Amadoro, G, De Angelis, R, Calissano, P, Passananti, C. (2003) Rb binding protein Che-1 interacts with Tau in cerebellar granule neurons. Modulation during neuronal apoptosis. *Mol Cell Neurosci, 24*, 1038-1050.

Becker, EB & Bonni, A. (2004). Cell cycle regulation of neuronal apoptosis in development and disease. *Prog Neurobiol, 72*, 1-25.

Biswas, SC, Liu, DX, Greene, LA. (2005). Bim is a direct target of a neuronal E2F-dependent apoptotic pathway. *J Neurosci, 25,* 8349-8358.

Bowser, R & Smith, MA. (2002). Cell cycle proteins in Alzheimer's disease: plenty of wheels but no cycle. J Alzheimers Dis, 4, 249-254.

Busser, J, Geldmacher, DS & Herrup, K. (1998). Ectopic cell cycle proteins predict the sites of neuronal cell death in Alzheimer's disease brain. *J Neurosci, .18:*2801-2807.

Cash, AD, Perry, G, Ogawa, O, Raina, AK, Zhu, X & Smith MA. (2002). Is Alzheimer's disease a mitochondrial disorder? *Neuroscientist, 8*: 489-496.

Cernak, I, Stoica, B, Byrnes, KR, Di Giovanni, S & Faden AI. (2005). Role of the cell cycle in the pathobiology of central nervous system trauma. *Cell Cycle, 4,* 1286-1293.

Chong, ZZ, Li, F & Maiese, K. (2005). Stress in the brain: novel cellular mechanisms of injury linked to Alzheimer's disease. *Brain Res Brain Res Rev, 49*, 1-21.

Cooper-Kuhn, CM, Vroemen, M, Brown J, Ye H, Thompson MA, Winkler J & Kuhn HG. (2002) Impaired adult neurogenesis in mice lacking the transcription factor E2F1. *Mol Cell Neurosci,* 21, 312-323.

Copani, A, Condorelli, F, Caruso, A, Vancheri, C, Sala, A, Giuffrida Stella, AM, Canonico, PL, Nicoletti, F, Sortino, MA. (1999). Mitotic signaling by beta-amyloid causes neuronal death. *FASEB J., 13*, 2225-2234.

Copani, A, Sortino, MA, Caricasole, A, Chiechio, S, Chisari, M, Battaglia, G, Giuffrida-Stella, AM, Vancheri, C & Nicoletti, F. (2002). Erratic expression of DNA polymerases by beta-amyloid causes neuronal death. *FASEB J, 16*, 2006-2008.

Copani, A, Uberti, D, Sortino, MA, Bruno, V, Nicoletti, F & Memo M. (2001). Activation of cell-cycle-associated proteins in neuronal death: a mandatory or dispensable path? *Trends Neurosc, 24*, 25-31.

Cregan, SP, Arbour, NA, Maclaurin, JG, Callaghan, SM, Fortin, A, Cheung, EC, Guberman, DS, Park, DS & Slack RS. (2004). p53 activation domain 1 is essential for PUMA upregulation and p53-mediated neuronal cell death. *J Neurosci, 24*, 10003-10012.

Dawson, VL & Dawson, TM. (2004). Deadly conversations: nuclear-mitochondrial cross-talk. *J Bioenerg Biomembr, 36,*287-94.

Dimova, DK, Dyson, NJ. (2005). The E2F transcriptional network: old acquaintances with new faces. *Oncogene, 24,* 2810-2826.

Eischen, CM, Packham, G, Nip, J, Fee, BE, Hiebert, SW, Zambetti, GP & Cleveland JL. (2001). Bcl-2 is an apoptotic target suppressed by both c-Myc and E2F1. *Oncogene, 20,* 6983-6993.

Fortin, A, MacLaurin, JG, Arbour, N, Cregan, SP, Kushwaha, N, Callaghan, SM, Park, DS, Albert, PR & Slack, RS. (2004).The proapoptotic gene SIVA is a direct transcriptional target for the tumor suppressors p53 and E2F1. *J Biol Chem, 279,* 28706-28714.

Francisconi, S, Codenotti, M, Ferrari-Toninelli, G, Uberti D & Memo M. (2005). Preservation of DNA integrity and neuronal degeneration. *Brain Res Brain Res Rev, 48,* 347-351.

Frappart, PO, Tong, WM, Demuth, I, Radovanovic, I, Herceg, Z, Aguzzi, A, Digweed, M & Wang, ZQ. (2005) An essential function for NBS1 in the prevention of ataxia and cerebellar defects. *Nat Med.* 11, 538-544.

Friedlander, R.M. (2003). Apoptosis and caspases in neurodegenerative diseases. *New England J Med. 3, 348,* 1365-1375.

Furukawa, Y, Nishimura, N, Furukawa, Y, Satoh, M, Endo, H, Iwase, S, Yamada, H, Matsuda, M, Kano, Y & Nakamura M. (2002). Apaf-1 is a mediator of E2F1-induced apoptosis. *J Biol Chem, 277,* 39760-39768.

Garcia-Calvo, M, Peterson, EP, Leiting, B, Ruel, R, Nicholson, DW & Thornberry, NA. (1998). Inhibition of Human Caspases by Peptide-based and Macromolecular Inhibitors. *J Biol Chem, 273*, 32608-32613.

Gastard, MC, Troncoso, JC & Koliatsos, V.E. (2003). Caspase activation in the limbic cortex of subjects with early Alzheimer's disease. *Ann Neurol, 54*, 393-398.

Gendron, TF, Mealing, GA, Paris, J, Lou, A, Edwards, A, Hou, ST, MacManus, JP, Hakim, AM & Morley P. (2001). Attenuation of neurotoxicity in cortical cultures and hippocampal slices from E2F1 knockout mice. *J Neurochem, 78*, 316-324.

Giovanni, A, Keramaris, E, Morris, EJ, Hou, ST, O'Hare, M, Dyson, N, Robertson, GS, Slack, RS & Park DS. (2000). E2F1 mediates death of B-amyloid-treated cortical neurons in a manner independent of p53 and dependent on Bax and caspase 3. *J Biol Chem, 2000, 275,* 11553-11560.

Giovanni, A, Wirtz-Brugger, F, Keramaris, E, Slack, R & Park DS. (1999) Involvement of cell cycle elements, cyclin-dependent kinases, pRb, and E2F x DP, in B-amyloid-induced neuronal death. *J Biol Chem. 1999, 274,* 19011-19016.

Herrup, K & Arendt, T. (2002). Re-expression of cell cycle proteins induces neuronal cell death during Alzheimer's disease. *J Alzheimers Dis, 4,* 243-247.

Herrup, K, Neve, R, Ackerman, SL & Copani A. (2004). Divide and die: cell cycle events as triggers of nerve cell death. *J Neurosci 2004, 24,* 9232-9.

Hershko, T & Ginsberg, D. (2004). Up-regulation of Bcl-2 homology 3 (BH3)-only proteins by E2F1 mediates apoptosis. *J Biol Chem, 2004, 279,* 8627-86 34.

Hou, ST, Callaghan, D, Fournier, MC, Hill, I, Kang, L, Massie, B, Morley, P, Murray, C, Rasquinha, I, Slack R & MacManus JP. (2000). The transcription factor E2F1 modulates apoptosis of neurons. *J Neurochem, 75,* 91-100.

Hou, ST, Cowan, E, Dostanic, S, Rasquinha, I, Comas, T, Morley, P & MacManus, JP. (2001). Increased expression of the transcription factor E2F1 during dopamine-evoked, caspase-3-mediated apoptosis in rat cortical neurons. *Neurosci Lett, 306,* 153-156.

Hou, ST, Cowan, E, Walker, T, Ohan, N, Dove, M, Rasqinha, I & MacManus JP. (2001). The transcription factor E2F1 promotes dopamine-evoked neuronal apoptosis by a mechanism independent of transcriptional activation. *J Neurochem, 78,* 287-97.

Irwin, M, Marin, MC, Phillips, AC, Seelan, RS, Smith, DI, Liu, W, Flores, ER, Tsai, KY, Jacks, T, Vousden, KH & Kaelin, WG. (2000). Role for the p53 homologue p73 in E2F1-induced apoptosis. *Nature, 407,* 645-648.

Jacobsen, E, Beach, T, Shen, Y, Li, R & Chang Y. (2004). Deficency of the Mre11 repair complex in Alzheimer's diseases brains. *Mol Brain Res, 128,* 1-7.

Jorda, EG, Verdaguer, E, Canudas, AM, Jimenez, A, Bruna, A, Caelles, C, Bravo, R, Escubedo, E, Pubill, D, Camarasa, J, Pallas, M & Camins A. Neuroprotective action of flavopiridol, a cyclin-dependent kinase inhibitor, in colchicine-induced apoptosis. *Neuropharmacology, 2003, 45,* 672-683.

Jordan-Sciutto, K, Rhodes, J & Bowser R. (2001). Altered subcellular distribution of transcriptional regulators in response to Abeta peptide and during Alzheimer's disease. *Mech Ageing Dev, 123,* 11-20.

Jordan-Sciutto, KL, Dorsey, R, Chalovich, EM, Hammond, RR & Achim CL. (2003). Expression patterns of retinoblastoma protein in Parkinson disease. *J Neuropathol Exp Neurol, 62,* 68-74.

Jordan-Sciutto, KL, Malaiyandi, LM & Bowser, R. (2002) Altered distribution of cell cycle transcriptional regulators during Alzheimer disease. *J Neuropathol Exp Neuro, 2002, 61,* 358-67.

Jordan-Sciutto, KL, Morgan, K & Bowser, R. (1999). Increased Cyclin G1 Immunoreactivity During Alzheimer's Disease. *J Alzheimers Dis, 1,* 409-417.

Jordan-Sciutto, KL, Wang, G, Murphey-Corb, M & Wiley, CA. (2002). Cell cycle proteins exhibit altered expression patterns in lentiviral-associated encephalitis. *J Neurosci, 22,* 2185-2195.

Jordan-Sciutto, KL, Wang, G, Murphy-Corb, M Wiley CA. (2000). Induction of cell-cycle regulators in simian immunodeficiency virus encephalitis. *Am J Pathol, 2000, 157,* 497-507.

Katchanov, J, Harms, C, Gertz, K, Hauck, L, Waeber, C, Hirt, L, Priller, J, von Harsdorf, R, Bruck, W, Hortnagl, H, Dirnagl, U, Bhide, PG & Endres M. (2000). Mild cerebral ischemia induces loss of cyclin-dependent kinase inhibitors and activation of cell cycle machinery before delayed neuronal cell death. *J Neurosci. 2001, 21,* 5045-53.

Kerr, JFR. (2002). History of the events leading to the formulation of the apoptosis concept. *Toxicology , 181-182,* 471-474.

Klein, JA, Ackerman, SL. (2003). Oxidative stress, cell cycle, and neurodegeneration. *J Clin Invest, 111*, 785-93.

Klein, JA, Longo-Guess, CM, Rossmann, MP, Seburn, KL, Hurd, RE, Frankel, WN, Bronson, RT, Ackerman, SL. (2002). The harlequin mouse mutation downregulates apoptosis-inducing factor. Nature. 419:367-374.

Konishi, Y, Bonni, A. (2003). The E2F-Cdc2 cell-cycle pathway specifically mediates activity deprivation-induced apoptosis of postmitotic neurons. *J Neurosci, 23*, 1649-1658.

Kroemer, G, Dallaporta, B & Resche-Rigon, M. (1998). The mitochondrial death/life regulator in apoptosis and necrosis. *Annu Rev Physiol, 60,* 619-42.

Kruman, II. (2004) Why do neurons enter the cell cycle? *Cell Cycle, 3*, 769-73.

Kruman II, Wersto RP, Cardozo-Pelaez, F, Smilenov, L, Chan, SL, Chrest, FJ, Emokpae, R Jr, Gorospe, M, Mattson, MP. (2004) Cell cycle activation linked to neuronal cell death initiated by DNA damage. *Neuron, 41,* 549-561.

Koulich, E, Nguyen, T, Johnson, K, Giardina, C & D'mello S. (2001). NF-kappaB is involved in the survival of cerebellar granule neurons: association of IkappaBbeta [correction of Ikappabeta] phosphorylation with cell survival. *J Neurochem, 76*, 1188-1198.

Lim, AC & Qi RZ. (2003). Cyclin-dependent kinases in neural development and degeneration. *J Alzheimers Dis, 5*, 329-335.

Liu, DX, & Greene, LA. (2001). Regulation of neuronal survival and death by E2F-dependent gene repression and derepression. *Neuron, 32*, 425-438.

Liu, DX, Biswas, SC & Greene LA. (2004). B-myb and C-myb play required roles in neuronal apoptosis evoked by nerve growth factor deprivation and DNA damage. *J Neurosci, 24*, 8720-8725.

Lockshin, R.A.& Zakari, Z. (2004). Apoptosis,autophagy and more. *Int J Biochem Cell Biol, 36,* 2405-2419.

Love, S. (2003). Neuronal expression of cell cycle-related proteins after brain ischaemia in man. *Neurosci Lett, 353,* 29-32.

Luo, Y, Hattori, A, Munoz, J, Qin, ZH & Roth GS. (1999) Intrastriatal dopamine injection induces apoptosis through oxidation-involved activation of transcription factors AP-1 and NF-kappaB in rats. *Mol Pharmacol, 56*, 254-264.

Maccioni, R.B., Munoz J.P. & Barbeito L. (2001). The molecular bases of Alzheimer's disease and other neurodegenerative disorders. *Arch Med Res , 32*, 367-381.

MacManus, JP, Jian, M, Preston, E, Rasquinha, I, Webster, J & Zurakowski B. (2003). Absence of the transcription factor E2F1 attenuates brain injury and improves behavior after focal ischemia in mice. *J Cereb Blood Flow Metab, 23,* 1020-1028.

MacManus, JP, Koch, CJ, Jian, M, Walker, T & Zurakowski, B. (1999). Decreased brain infarct following focal ischemia in mice lacking the transcription factor E2F1. *Neuroreport, 10,*: 2711-2714.

Martin-Romero, FJ, Santiago-Josefat, B, Correa-Bordes, J, Gutierrez-Merino, C & Fernandez-Salguero, P. (2000). Potassium-induced apoptosis in rat cerebellar granule cells involves cell-cycle blockade at the G1/S transition. *J Mol Neurosci*, 15, 155-65.

McMurray, CT. (2005). To die or not to die: DNA repair in neurons. *Mutat Res. 2005,* 577:260-274.

Mirjany, M, Ho, L & Pasinetti GM. (2002). Role of cyclooxygenase-2 in neuronal cell cycle activity and glutamate-mediated excitotoxicity. *J Pharmacol Exp Ther, 301*, 494-500.

Motonaga, K, Itoh, M, Hirayama, A, Hirano, S, Becker, LE, Goto, Y & Takashima S. (2001). Up-regulation of E2F1 in Down's syndrome brain exhibiting neuropathological features of Alzheimer-type dementia. *Brain Res, 905*, 250-253.

Nagy, Z, Esiri, MM & Smith AD. (1998). The cell division cycle and the pathophysiology of Alzheimer's disease. *Neuroscience, 87*, 731-739.

Nahle, Z, Polakoff, J, Davuluri, RV, McCurrach, ME, Jacobson, MD, Narita, M, Zhang, MQ, Lazebnik, Y, Bar-Sagi, D & Lowe SW. (2002). Direct coupling of the cell cycle and cell death machinery by E2F. *Nat Cell Biol., 4*, 859-864.

Neema, M, Navarro-Quiroga, I, Chechlacz, M, Gilliams-Francis, K, Liu, J, Lamonica, K, Lin, SL & Naegele JR. (2005). DNA damage and nonhomologous end joining in excitotoxicity: neuroprotective role of DNA-PKcs in kainic acid-induced seizures. *Hippocampus 15*: 1057-1071.

Neve, RL & McPhie DL. (2006). The cell cycle as a therapeutic target for Alzheimer's disease. *Pharmacol Ther, 111*, 99-113.

Nicotera, P. & Leist,M. (1997). Energy supply and the shape of death in neurons and lymphoid cells. *Cell Death and Differ, 4*, 435-442.

Nishioka, N & Arnold, SE. (2004). Evidence for oxidative DNA damage in the hippocampus of elderly patients with chronic schizophrenia. *Am J Geriatr Psychiatry, 12,* 167-75.

O'Hare, MJ, Hou, ST, Morris, EJ, Cregan, SP, Xu, Q, Slack, RS & Park DS. (2000). Induction and modulation of cerebellar granule neuron death by E2F1. *J Biol Chem, 275*, 25358-25264.

Osuga, H, Osuga, S, Wang, F, Fetni, R, Hogan, MJ, Slack, RS, Hakim, AM, Ikeda, JE & Park DS. (2000). Cyclin-dependent kinases as a therapeutic target for stroke. *Proc Natl Acad Sci U S A, 97*, 10254-10259.

Otsuka, Y, Tanaka, T, Uchida, D, Noguchi, Y, Saeki, N, Saito, Y & Tatsuno I. (2004). Roles of cyclin-dependent kinase 4 and p53 in neuronal cell death induced by doxorubicin on cerebellar granule neurons in mouse. *Neurosci Lett, 365*, 180-185.

Padmanabhan, J, Park, DS, Greene, LA, Shelanski, ML. (1999). Role of cell cycle regulatory proteins in cerebellar granule neuron apoptosis. *J Neurosci, 19*, 8747-8756.

Pallas, M, Verdaguer, E, Jorda, EG, Jimenez, A, Canudas, AM & Camins A. (2005). Flavopiridol: an antitumor drug with potential application in the treatment of neurodegenerative diseases. *Med Hypotheses, 64,* 120-123.

Park, DS, Morris, EJ, Greene, LA & Geller, HM. (1997). G1/S cell cycle blockers and inhibitors of cyclin-dependent kinases suppresscamptothecin-induced neuronal apoptosis. *J Neurosci, 17,* 1256-1270.

Park, DS, Morris, EJ, Stefanis, L, Troy, CM, Shelanski, ML, Geller, HM & Greene LA. (1998). Multiple pathways of neuronal death induced by DNA-damaging agents, NGF deprivation, and oxidative stress. *J Neurosci,18,* 830-840.

Park, DS, Obeidat, A, Giovanni, A & Greene LA. (2000) Cell cycle regulators in neuronal death evoked by excitotoxic stress: implications for neurodegeneration and its treatment. *Neurobiol Aging, 21,* 771-781.

Phillips, AC, Ernst, MK, Bates, S, Rice, NR & Vousden KH. (1999). E2F1 potentiates cell death by blocking antiapoptotic signaling pathways. *Mol Cell, 4*, 771-781.

Putzer, BM, Tuve, S, Tannapfel, A, Stiewe, T. (2003) Increased DeltaN-p73 expression in tumors by upregulation of the E2F1-regulated, TA-promoter-derived DeltaN'-p73 transcript. *Cell Death Differ. 10,* 612-624.

Ranganathan, S, Bowser, R. (2003) Alterations in G(1) to S phase cell-cycle regulators during amyotrophic lateral sclerosis. *Am J Pathol, 162*, 823-835.

Rashidian, J, Iyirhiaro, G, Aleyasin, H, Rios, M, Vincent, I, Callaghan, S, Bland, RJ, Slack RS, During MJ & Park DS. (2005). Multiple cyclin-dependent kinases signals are critical mediators of ischemia/hypoxic neuronal death in vitro and in vivo. *Proc Natl Acad Sci U S A, 102,* 14080-14085.

Reed, J.C. (2001). Apoptosis-regulating proteins as targets for drug discovery. *Trends Mol Med, 7,* 314-319

Shimohama S. (2000). Apoptosis in Alzheimer's disease. An update. *Apoptosis, 5*, 9-16.

Smith, RA, Walker, T, Xie, X & Hou ST. (2003). Involvement of the transcription factor E2F1/Rb in kainic acid-induced death of murine cerebellar granule cells. *Brain Res Mol Brain Res, 116*, 70-79.

Smith, PD, O'Hare, MJ & Park DS. (2004). Emerging pathogenic role for cyclin dependent kinases in neurodegeneration. *Cell Cycle.3*, 289-291.

Stanelle, J & Putzer, BM. (2006). E2F1-induced apoptosis: turning killers into therapeutics. *Trends Mol Med, 12,* 177-185.

Stanelle, J, Tu-Rapp, H & Putzer, BM. (2005). A novel mitochondrial protein DIP mediates E2F1-induced apoptosis independently of p53. *Cell Death Differ, 12*, 347-57.

Stefanis, L, Park, DS, Friedman, WJ & Greene LA. (1999). Caspase-dependent and -independent death of camptothecin-treated embryonic cortical neurons. *J Neurosci, 19,* 6235-47.

Strachan, GD, Koike, MA, Siman, R, Hall, DJ & Jordan-Sciutto, KL. (2005). E2F1 induces cell death, calpain activation, and MDMX degradation in a transcription independent manner implicating a novel role for E2F1 in neuronal loss in SIV encephalitis. *J Cell Biochem, 96,* 728-740.

Troy, C.M. & Salvesen, G.S. (2002). Caspases on the brain. *J Neurosci Res, 69,* 145-150.

Ueberham U & Arendt T. (2005). The expression of cell cycle proteins in neurons and its relevance for Alzheimer's disease. *Curr Drug Targets CNS Neurol Disord., 4*, 293-306.

Van Loo, G., Saelens; X., van Gurp, M., MacFarlane, M, Martin, S.J. & Vandenabeele, P. (2002). The role of mitochondrial factors in apoptosis: a Russian roulette with more than one bullet. *Cell Death Diff, 2002*, 9, 1031-1042.

Verdaguer E, Garcia-Jorda E, Canudas AM, Dominguez E, Jimenez A, Pubill D, Escubedo E, Camarasa, J, Pallas M, Camins A. (2002). Kainic acid-induced apoptosis in cerebellar granule neurons: an attempt at cell cycle re-entry. *Neuroreport, 13,* 413-416.

Verdaguer, E, Jorda, EG, Alvira, D, Jimenez, A, Canudas, AM, Folch, J, Rimbau, V, Pallas, M, Camins, A. (2005). Inhibition of multiple pathways accounts for the antiapoptotic effects of flavopiridol on potassium withdrawal-induced apoptosis in neurons. *J Mol Neurosci, 26,* 71-84.

Verdaguer, E, Jimenez, A, Canudas, AM, Jorda, EG, Sureda, FX, Pallas, M & Camins A. (2004). Inhibition of cell cycle pathway by flavopiridol promotes survival of cerebellar granule cells after an excitotoxic treatment. *J Pharmacol Exp Ther, 308,* 609-616.

Wang, H, Yu, SW, Koh, DW, Lew, J, Coombs, C, Bowers, W, Federoff, HJ, Poirier, GG, Dawson, TM & Dawson VL. (2004). Apoptosis-inducing factor substitutes for caspase executioners in NMDA-triggered excitotoxic neuronal death. *J Neurosci. 24,* 10963-10973.

Wang B, Liu K, Lin FT, Lin WC. (2004) A role for 14-3-3 tau in E2F1 stabilization and DNA damage-induced apoptosis. *J Biol Chem. 279,* 54140-52.

Wang, F, Corbett, D, Osuga, H, Osuga, S, Ikeda, JE, Slack, RS, Hogan, MJ, Hakim, AM & Park, DS. (2002). Inhibition of cyclin-dependent kinases improves CA1 neuronal survival and behavioral performance after global ischemia in the rat. *J Cereb Blood Flow Metab, 22,* 171-82.

Wang, H, Yu, SW, Koh, DW, Lew, J, Coombs, C, Bowers, W, Federoff, HJ, Poirier, GG, Dawson, TM & Dawson, VL. (2004). Apoptosis-inducing factor substitutes for caspase executioners in NMDA-triggered excitotoxic neuronal death. *J Neurosci, 24,* 10963-10973

Webber, KM, Raina, AK, Marlatt, MW, Zhu, X, Prat, MI, Morelli, L, Casadesus, G, Perry, G & Smith, MA. (2005). The cell cycle in Alzheimer disease: a unique target for neuropharmacology. *Mech Ageing Dev,* 126, 1019-1025.

Wen,Y, Yang, S, Liu, R, Simpkins, JW. (2005). Cell-cycle regulators are involved in transient cerebral ischemia induced neuronal apoptosis in female rats. *FEBS Lett. 579,* 4591-4599.

Wu Chen, R, Zhang, Y, Rose, ME & Graham, SH. (2004). Cyclooxygenase-2 activity contributes to neuronal expression of cyclin D1 after anoxia/ischemia in vitro and in vivo. *Brain Res Mol Brain Res, 132,* 31-37.

Xie, W, Jiang, P, Miao, L, Zhao, Y, Zhimin, Z, Qing, L, Zhu, WG & Wu M. (2006). Novel link between E2F1 and Smac/DIABLO: proapoptotic Smac/DIABLO is transcriptionally upregulated by E2F1. *Nucleic Acids Res, 34,* 2046-2255.

Yang, Y, Geldmacher, DS & Herrup K. (2001). DNA replication precedes neuronal cell death in Alzheimer's disease. *J Neurosci, 21,* 2661-2668.

Yang, Y, Mufson, EJ, Herrup, K. (2003). Neuronal cell death is preceded by cell cycle events at all stages of Alzheimer's disease. *J Neurosci, 23,* 2557-2563.

Yu, X, Caltagarone, J, Smith, MA, & Bowser R. (2005). DNA damage induces cdk2 protein levels and histone H2B phosphorylation in SH-SY5Y neuroblastoma cells. *J Alzheimers Dis 8,* 7-21.

Zhang, M, Li, J, Chakrabarty, P, Bu, B & Vincent, I. (2004). Cyclin-dependent kinase inhibitors attenuate protein hyperphosphorylation, cytoskeletal lesion formation, and motor defects in Niemann-Pick Type C mice. *Am J Pathol, 165,* 843-53.

Zhu, X, Raina, AK, Perry, G & Smith, MA. (2004). Alzheimer's disease: the two-hit hypothesis. *Lancet Neurol.* 3, 219-226.

In: Neuronal Cell Apoptosis Research
Editor: Relif J. Shandling, pp. 97-120

ISBN 1-60021-456-8


Chapter 4

# DNA Damage Response and Apoptosis of Postmitotic Neurons

*Inna I. Kruman*[1,*] *and Elena I. Schwartz*[2]
[1]Sun Health Research Institute, 1015 West Santa Fe Drive, Sun City, AZ 85351;
[2]Department of Pathology and Oncology, Georgetown University Medical Center, 3900 Reservoir RD NW, Washington, DC 20057.

## Abstract

Programmed cell death or apoptosis is a relevant process in the physiology and pathology of the nervous system. Apoptosis is an organized form of cell death which is triggered by different factors including DNA damage. A growing body of evidence suggests that DNA damage and genomic instability are involved in neuronal abnormalities and may play a central role in neurodegeneration. DNA strand breaks and DNA lesions have been reported in Parkinson's and Alzheimer's diseases and as an early event after reperfusion of ischemic brain. DNA damage has been found to activate a cell death program in terminally differentiated postmitotic neurons. Since the genome is continuously damaged by a variety of endogenous and exogenous agents and the majority of DNA damage is produced by oxyradicals, generated by normal aerobic metabolism, neurons are particularly susceptible to DNA damage due to the high rate of oxidative metabolism. To maintain genomic integrity, cells are equipped with special defense mechanism, DNA damage response, to remove DNA damage by DNA repair pathways or eliminate damaged cells via apoptosis. Generally, differentiated cells, like neurons, are deficient in DNA repair and more vulnerable to DNA damage-initiated apoptosis. For example, neurons are more vulnerable than astrocytes to DNA-damaging conditions such as ionizing radiation. Breast cancer patients receiving hemotherapy commonly experience long-lasting cognitive impairment. It is known that radiotherapy

* Correspondence concerning this article should be addressed to Correspondence to: Inna Kruman, Ph.D.; Sun Health Research Institute, 1015 West Santa Fe Drive, Sun City, AZ 85351; Tel.: 623-876-5607; FAX 623- 876-5695; Email: inna.kruman@sunhealth.org.

may cause CNS toxicity. DNA damaging agents including γ-irradiation induce neuronal apoptosis *in vitro*, suggesting the direct adverse effect of these DNA damaging agents on neurons. The importance of DNA repair for neuronal survival is illustrated by disorders observed in patients with hereditary DNA repair abnormalities. These disorders combine the predisposition to cancer with progressive neurodegeneration. Although indirect evidence suggests that DNA damage and repair mechanisms play critical roles in neuronal survival, the pathways involved are poorly understood. Recently, we have found that cell cycle activation is essential for DNA damage-induced neuronal apoptosis which suggests that the cell cycle machinery is a critical element of the DNA damage response not only in cycling but also in quiescent cells. Here, we discuss the DNA damage response in postmitotic neurons and possible mechanisms by which neurons are forced to apoptosis versus DNA repair thereby controlling cell fate. Elucidation of these mechanisms promises to provide multiple points of therapeutic intervention in neurodegenerative diseases.

## INTRODUCTION

Neuronal cell death is both a normal developmental process and the result of trauma or degenerative disorders. During development, a significant portion of neurons dies (Oppenheim, 1991), and this death is a part of normal process leading to appropriate matching of pre- and postsynaptic elements Greene et al., 2004). On the other hand, loss of neurons may be a part of the pathological process. Understanding the mechanisms of neuronal death, thus provides insights about nervous system development and anunderstanding of the pathogenesis of neurodegeneration.

Apoptosis is a regulated form of cell death that is often likened to cellular suicide. Unlike necrosis, apoptosis is an active ATP-dependent biochemical process which is regulated through transcriptional, as well as post-translational mechanisms and is not accompanied by inflammation. Apoptotic cells undergo shrinkage, membrane blebbing, chromatin condensation, DNA fragmentation, and cellular disintegration followed by phagocytosis (Wyllie et al., 1980). Many neurons die by apoptosis during early CNS development (Burek and Oppenheim, 1996). In addition, apoptosis contributes to neuronal loss in a number of acute and chronic neurological diseases including stroke, Alzheimer's disease (AD), and amyotrophic lateral sclerosis (ALS) (Friedlander, 2003; Honig and Rosenberg, 2000; Przedborski and Vila, 2003 and Yuan and Yankner, 2000). In general, neuronal apoptosis in these disorders is thought to be caused by toxic insults such as free-radical generation (Klein and Ackerman, 2003). The intracellular end point of many neurotoxic stimuli including excitotoxicity and exposure to amyloid β peptide (Aβ, an essential feature of Alzheimer's disease) is oxidative stress (Butterfield, 2002 and Lewen et al., 2000). Oxidative stress in the form of increased reactive oxygen species (ROS) appears to be a common apoptotic trigger leading to neurodegeneration (Carri et al., 2003; Klein and Ackerman, 2003). ROS include the superoxide anion, hydrogen peroxide, hydroxyl radical, peroxynitrite and lipid peroxides. ROS are constantly produced within the cell, in particular via mitochondrial oxidative metabolism and pathological processes such as inflammation. ROS are also secondary messengers in specific signaling pathways, generated by plasma membrane oxidases in response to growth factors and cytokines (Mahmoudi et al., 2006); they are and therefore, involved in normal cell signaling. The cell has a number of mechanisms of protection against oxidative damage (Offord et al., 2000), including direct interaction with antioxidants such as α-tocopherol, ascorbic acid and glutathione. In general, the balance between ROS levels and the activity of these defense mechanisms determines the degree of oxidative stress encountered by the cell. Neurons are highly susceptible to oxidative stress due to their high rate of oxidative metabolism and low level of antioxidant enzymes (Brook, 2000). ROS are thought to lead to apoptosis by the destruction of lipids, proteins, and nucleic acids (Kannan and Jain, 2000). DNA is perhaps the major target of ROS. In many neurodegenerative disorders, DNA damage is thought to contribute to neuronal loss. DNA damage has been found in chronic neurodegenerative conditions such as Parkinson's (Alam et al., 1997; Jenner and Olanow, 1998) and Alzheimer's diseases (Mecocci et al., 1998; Lovell and Markesbery, 2001), as well as amyotrophic lateral sclerosis (Bogdanov et al., 2000).

## THE DNA DAMAGE RESPONSE

Living cells are continuously exposed to endogenous and environmental DNA-damaging agents. More than $10^4$ DNA-damaging events occur in each mammalian cell every day from spontaneous decay, replication errors and cellular metabolism alone (Foray et al., 2003). A major portion of DNA damage occurs via ROS (Williams and Jeffrey, 2000). Exogenous damage may be caused by ultraviolet (UV) radiation from the sun, x- and gamma rays, hydrolysis or thermal disruption, certain plant toxins, such as botulin, or human-made mutagenic chemicals acting as DNA intercalating agents (Jeggo and Lobrich, 2006). If not repaired, some of these lesions are cytotoxic or mutagenic. A cell that has accumulated a large amount of DNA damage, or one that no longer effectively repairs damage, can enter one of three possible states:an irreversible state of dormancy, known as senescence, apoptosis, or mutagenesis and carcinogenesis (Figure1). The protection of genomic integrity, therefore, is a major challenge for living cells. To cope with the deadly consequences of DNA lesions, that interfere with essential DNA-dependent processes such as transcription and replication, cells are equipped with an efficient defense mechanism termed the DNA damage response. The function of the DNA damage response is to eliminate DNA damage through DNA repair or to remove cells with incurred DNA damage by means of apoptosis. DNA damage response mechanisms encompass pathways of DNA repair, cell cycle checkpoint arrest and apoptosis. Thus, two systems are essential for genome integrity, DNA repair and apoptosis (Shiloh, 2006). Cells defective in DNA repair tend to accumulate excess DNA damage, while cells defective in apoptotic process, tend to survive with excess DNA damage. In this regard, the cellular response to DNA damage is crucial for maintaining homeostasis, preventing the development of cancer and neurodegeneration. Defects in this response lead to severe "genomic instability syndromes" (Brooks, 2002). The many types of DNA lesions are rapidly detected, with subsequent activation of a web of signaling pathways which together compose the DNA damage response, culminating in the activation of cell-cycle checkpoints and appropriate DNA repair pathways, or the initiation of apoptotic programs.

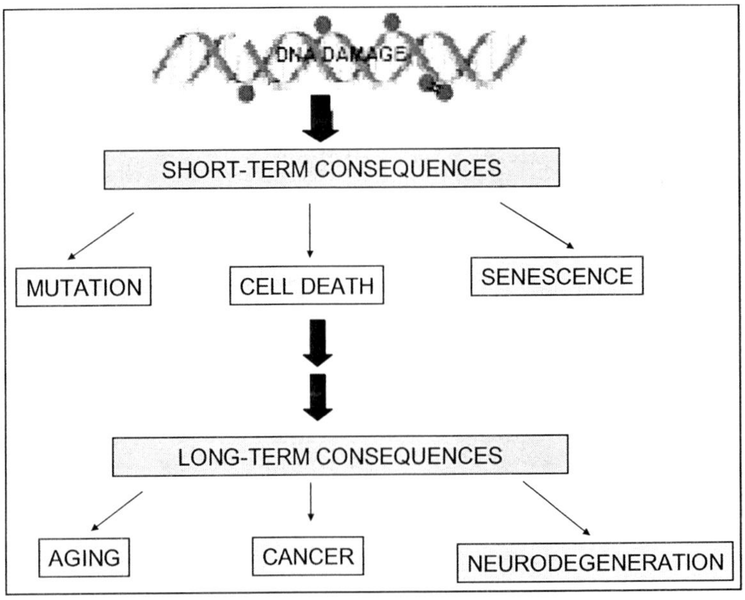


Figure 1. The consequences of non-repaired DNA damage.

## DNA Damage Check Points

Checkpoints tightly control progress throughout the cell cycle. Cells may be arrested at any of the checkpoints, and the DNA will be repaired, or alternately cells may die by apoptosis (Sancar et al., 2004). The latter is a mechanism to ensure that non-repairable DNA modifications are not passed on to the progeny of a damaged cell. The same proteins involved in regulating the orderly progression through the cell cycle are also involved in the checkpoint responses (Sancar et al., 2004). The process of DNA replication itself in addition to the usual DNA damaging insults, adds intrinsic risks, such as base misincorporation errors and stalled replication forks, which demand an immediate response from the checkpoint machinery to preserve genomic integrity. If left uncorrected, the errors may lead to the nucleotide sequence alterations and chromosomal rearrangements usually associated with cellular transformation (Kunkel and Bebenek, 2000). The integrity of the S-phase is monitored mainly via two phosphoinositide 3-kinase (PI3)-related kinases ATM (ataxia telangiectasia mutated) and ATR (ATM and Rad 3-related), which play critical roles in early signal transmission (Tibbetts et al, 2000; Abraham, 2003). These two serine/threonine kinases play a central role in signaling DNA damage (Abraham, 2003). $G_1$- to S-phase transition requires the coordinated action of many proteins including members of the E2F family of transcription factors (Harbour and Dean 2000). E2F activity is required for the transcriptional induction of many genes required for cell cycle progression including cyclins, cyclin dependent kinases (CDKs) and enzymes involved in DNA replication. In addition, E2F functions in apoptosis following DNA damage and regulates expression of pro-apoptotic genes including *apaf-1* and several caspases (DeGregori, 2002). ATM and ATR are thought to be involved in the transmission of signals from stalled replication forks through a multitude of signaling cascades. It has been found that E2F-1 protein levels, DNA binding activity and transcriptional activity are increased following exposure to UV light (O'Connor and Lu, 2000; Blattner et al et al., 1999) or other DNA-damaging agents (Meng et al., 1999). Thus, DNA-damage may induce apoptosis through E2F-1. In turn, the association of this apoptosis with S-phase may result from the fact that E2F-1 is in its free form (i.e. not complexed with retinoblastoma protein Rb (pRb) and thus most active during this phase of the cell-cycle (Harbour and Dean, 2000). The pRb and its related proteins (p107, p130) are thought to act as a checkpoint for G1/S transition in proliferating cells (Niida and Nakanishi, 2006). In its inactive state, pRb is non-phosphorylated and is believed to inhibit cell cycle progression by repressing E2F transcription factors. Upon mitogenic stimulation, pRb is phosphorylated by active cyclin D/CDK4/6 complexes and released from E2F, which results in the derepression and transactivation of E2F-target genes required for S phase entry (Harbour and Dean, 2000; Liu and Greene, 2001). However, there is evidence of the involvement of pRb, possibly activated and recruited to E2F-binding sites, in response to DNA damage (Knudsen et al., 2000). In addition, several genes involved in DNA repair were shown to be regulated by E2F. Deregulation of E2F upregulates the expression of genes involved in the DNA-damage response, e.g. ATM, BRCA1 and RAD51 (Polager et al., 2002; Ren et al., 2002; Berkovich and Ginsberg, 2003). The E2F1 protein has been shown to be stabilized and its levels to rise in response to DNA damage (Blattner et al., 1999; Lin et al., 2001). These findings support a role for E2F transcription factors in the DNA damage response. E2F regulation of DNA repair-associated genes at G1/S transition is consistent with the view that DNA replication is an error-prone process and that DNA repair is tightly linked to DNA synthesis (Boudsocq et al., 2002; Ren et al., 2002). Thus, the DNA damage response, apoptosis and the cell cycle share some common participants such as pRb, E2F, and ATM/ATR (King and Cidlowski, 1995; Stevaux and Dyson, 2002).

The activation of cell cycle checkpoints arrests the cell cycle, while the damage is assessed and repaired. DNA damage response involves alterations in numerous physiological processes such as gene expression and protein synthesis, degradation and trafficking (Shiloh, 2006). The DNA damage response is executed through a series of steps (Figure2). First, the DNA lesions are detected by sensor proteins. Damage to DNA modifies the spatial configuration of the helix and these modifications can be detected by the cell. Transducers initiate a signal transduction cascade that amplifies the signal and conveys the damage signal to downstream effectors. When damage is localized, specific DNA repair molecules, which are part of the whole DNA repair machinery, are recruited and bind at the site of damage, inducing other molecules to bind and form a complex that facilitates the actual DNA repair mechanism. The transducers might also be involved in the assembly of DNA-repair complexes at the sites of DNA damage (Iliakis et al., 2003; Shiloh, 2003; Niida and Nakanishi, 2006). The vast majority of DNA damage affects the primary structure of the double helix. The bases, which are the main component of the double helix, are chemically modified by themselves. These modifications can disrupt the regular helical structure of DNA by introducing non-native chemical bonds or bulky adducts that are not consistent with the standard double helix (Veglia et al., 2003).

There are four main types of chemical DNA damage that arise mainly from endogenous processes that generate ROS, especially superoxide and peroxides: oxidation of bases [e.g. 8-oxo-7,8-dihydroguanine (8-oxoG)] and generation of DNA strand interruptions from reactive

well-known mechanism is base excision repair (BER). In general, this mechanism requires removal of a damaged region by specialized nuclease systems and then DNA synthesis to fill the gap. BER involves five primary steps: (1) base excision, (2) AP site incision, (3) terminus "cleanup," (4) gap filling, and (5) nick ligation (Parikh et al., 1997). In the process of repair, corresponding DNA glycosylase removes a modified base from the DNA backbone, creating a transient abasic (AP) site (Nilsen et al., 1997; Slupphaug et al., 1991). AP endonuclease 1 (Ape1) cleaves the DNA backbone, which permits incorporation of a correct nucleotide by DNA polymerase β (β-pol) (Srivastava et al., 1998). Completion of BER is accomplished upon ligation (Dimitriadis et al., 1998).

## Single-Strand Break Repair

SSBs arise primarily from attack by ROS, whereas indirect SSBs are mainly normal intermediates of DNA BER, as a result of AP endonuclease activity and before they are "handed" to the next enzyme in the BER process, in a molecular relay. In addition to being the most common 'endogenous' lesion (thousands per cell per day) SSBs are also the most common lesions induced by exogenous genotoxins such as ionizing radiation, and alkylating agents (Caldecott, 2004). SSB repair contains four basic steps, beginning with DNA damage binding. SSB detection is largely achieved by the poly (ADP-ribose) polymerases, PARP-1 or PARP-2. A subset of SSBs arising during BER may also require detection or binding by PARPs. These proteins rapidly bind to and are activated by DNA strand breaks, and then covalently modify both themselves and other target proteins with poly (ADP-ribose).

## Double-Strand Break Repair

A type of DNA damage particularly hazardous to cells is a break to both strands in the double helix, double strand breaks (DSBs). DSBs are more common than was once expected and are estimated to occur at levels of 10 per cell per day in mammals. This is because they are caused not only by environmental sources, such as ionizing radiation and exposure to genotoxic compounds, but also, and more importantly, from endogenous sources, such as free radicals generated by oxidative respiration (Pfeiffer et al., 2000; Karanjawala et al., 2002). The stability of the mammalian genome therefore relies on efficient repair of DNA DSBs. Two mechanisms exist to repair this damage - non-homologous end-joining (NHEJ) and homologous recombination (HR) (Watson et al., 2004). The NHEJ pathway operates when the cell has not yet replicated the region of DNA on which the lesion has occurred. The process directly joins the two ends of the broken DNA strands without a template, losing sequence information in the process. Thus, this repair mechanism is necessarily mutagenic. However, if the cell is not dividing and has not replicated its DNA, the NHEJ pathway is the cell's only option (Wang et al., 2003). NHEJ involves DNA dependent protein kinase (DNA-PK), a holoenzyme consisting of a catalytic subunit (DNA-PKcs) and a DNA binding and regulatory subunit, Ku heterodimers (Ku 70 and Ku 80), DNA ligase-IV (LIG4), XRCC4 and Artemis (Lieber, 1999; Ma et al., 2002). Mice deficient in LIG4 (or XRCC4) exhibit a pleiotropic phenotype, including late embryonic lethality, cellular growth defects, ionizing-radiation sensitivity, and massive apoptosis of newly generated neurons (Gao et al., 1998; Barnes et al., 1998). Ku70- (Gu et al., 1997) and Ku80- (Nussenzweig et al., 1996) null mice exhibit a similar phenotype. The severe combined immune deficiency (SCID) mice which are deficient in DNA-PK-deficient exhibit severe immunodeficiency, and neurons are hypersensitive to DSB-induced apoptosis (Culmsee et al., 2001; Vemuri et al., 2001), suggesting the importance of NHEJ pathway. One of the very early events following DSBs is the phosphorylation of histone H2AX on serine 139 and simultaneously the autophosphorylation of Ser1981 of the nuclear protein kinase ataxia telangiectasia mutated (ATM) which is the primary DSB transducer (Sedelnikova et al., 2003; Bakkenist and Kastan, 2003). H2AX phosphorylation at serine 139 occurs at sites surrounding DSBs. Recent reports indicate that the dephosphorylation of γH2AX and dispersal of γH2AX foci in γ-irradiated cells correlate with DNA DSB repair (Rothkamm et al., 2003; MacPhail et al., 2003; Nazarov et al., 2003) and numerically correspond to DNA DSBs (Sedelnikova et al.,2002). The ATM function is critical for many aspects of the DSB response (Shiloh, 2003). After DSB induction, ATM is rapidly activated by autophosphorylation (Bakkenist and Kastan, 2003). Active ATM then phosphorylates different substrates, involved in the DNA damage response. An important mediator of cell-cycle checkpoints and damage-induced apoptosis is the tumor suppressor protein p53, which is activated by ATM via various posttranslational modifications (Meek, 2004) or by phosphorylating Mdm2, a ubiquitin ligase of p53, thereby targeting p53 for degradation (Stommel and Wahl, 204). In addition to its role in DNA NHEJ repair, DNA-PK plays important roles in the DNA damage response signaling (Jackson, 2002; Yang et al., 2003). The accumulating data point to an important role of p53 in the DNA-PK-mediated DNA damage signaling pathway (Achanta et al., 2001). Recent observations suggest that in contrast to ATM, DNA-PK is not involved in cell cycle checkpoint regulation through p53, but serves as an upstream regulator of the p53-mediated apoptosis pathway (Burma and Chen, 2004). On the other hand, DSB-induced apoptosis has been reported in p53-deficient cells (Lips and Kaina, 2001). Thus, the mechanisms by which DNA-PK and ATM work in DSB-induced signaling remain unclear. Another transcription factor that has been recently found to be a direct target of ATM is the $Ca^{2+}$/cAMP response element binding protein (CREB), which is involved in various cellular growth pathways (Shi et al., 2004). The loss of ATM activity is associated with the autosomal recessive disorder ataxia-telangiectasia (A-T), which is characterized by neuronal degeneration, severe neuromotor dysfunction, immunodeficiency, chromosomal fragility, a marked predisposition to cancer, and sensitivity to ionizing radiation (Shiloh, 2003).

## DNA Damage in the Nervous System

DNA damage is an important initiator of neuronal cell death and has also been implicated in neurodegenerative conditions. Given that the DNA of postmitotic neurons is particularly susceptible to oxidative stress because of the high rate of oxidative metabolism in the brain, and also due to the fact that postmitotic neurons may live for many decades, DNA damage could be a critical factor in ageing and the progression of some pathologies. DNA strand breaks have been reported in neurons after reperfusion of ischemic tissue, well in advance of DNA fragmentation caused by the apoptotic process (Tobita et al., 1995; Chen et al., 1997; Cui et al., 2000). DNA damage has also been found in various chronic neurodegenerative conditions (Alam et al., 1997; Jenner and Olanow, 1998; Mecocci et al., 1998; Bogdanov et al., 2000; Lovell and Markesbery, 2001). A two-fold higher incidence of single-strand breaks and other alkali-labile DNA lesions are detected in the cerebral cortex of AD versus control brains (Mullaart et al., 1990). In this regard, factors known to be relevant to DNA damage, such as poly (ADP-ribose) polymerase (PARP) and p53, have been implicated in a variety of neurodegenerative diseases (Cosi et al., 1997). Thus, neuronal DNA damage often occurs as a result of oxidative stress (Ratan et al. 1994; Park et al. 1998; Martin et al. 2003). According to our data, elevated levels of oxo8dG in brains of APP mutant transgenic mice, a mouse model of Alzheimer's disease, are correlated with increased levels of soluble Aβ1-42/Aβ1-40 (Kruman et al., 2004a). The amount of oxo8dG was found to be elevated in nuclear and mitochondrial DNA in neurons of diseased brain regions in patients with neurodegenerative disorders (Iida et al., 2002). The importance of DNA damage is illustrated further by the observation that neurological abnormalities are known to accompany defective DNA repair in various human syndromes such as ataxia telangiectasia and Cockayne syndrome (Rolig and McKinnon, 2000). Since neurons have a high metabolic rate and extensive exposure to oxidative stress, it is not surprising that defects in various branches of the DNA damage response lead to severe neurological demise (Brooks, 2002).

In further support, mice deficient in a variety of DNA damage repair pathways display aberrant neuronal loss and impaired neural development (Gao et al., 1998; Sugo et al., 2000), demonstrating connections between DNA damage and neurodegeneration. The importance of DNA repair for the nervous system is illustrated by the fact that many neurological abnormalities and progressive neurodegeneration in patients with hereditary diseases are associated with defects in DNA repair (Rolig and McKinnon, 2000). The nervous system most likely has a DNA repair system largely similar to that found in other tissues (Brooks, 2002; Biton et al., 2006). Terminally differentiated neurons are able to repair DNA damage, although the process is slower than in proliferating cells, and neurons are more prone to DNA damage-initiated apoptosis (Gobbel et al., 1998; Morris and Geller, 1996). A close link between DSB DNA damage and neurodegeneration, essential during nervous system development, appears evident from many pathological data and observations with mouse knockouts. Thus, DNA damage is an important initiator of neuronal cell death implicated in neurodegenerative conditions.

## DNA Damage Response in Neurons

Since neurons have an elevated rate of metabolism, particularly oxidative metabolism, which may result in accumulation of numerous lesions in the genome and may compromise transcription (Brooks, 2002), it would be logical to suggest that a mechanism exists to maintain the integrity of those genes needed for viable cell function. Neurons can repair DNA damage, but the process is slower than in proliferating cells, and neurons are known to be more prone to apoptosis (Gobbel et al., 1998). Upon terminal differentiation of neurons, there is a sharp decrease in global genomic repair (GGR). However, repair of active genes has been found to be maintained or even enhanced by differentiation (Ho and Hanawalt, 1991). Differentiated cells have been shown to repair transcribed genes using the specialized repair pathway termed transcription-coupled repair (TCR), which targets repair systems to transcribed genes (Nouspikel and Hanawalt, 2002). The mechanisms of this type of DNA repair are not yet fully understood. Several DNA repair systems, including base excision repair (BER) may be involved in TCR (Larsen et al., 2000 ).

## Role of Cell Cycle Reentry in the DNA Damage Response

The DNA damage response has been investigated mainly in proliferating cells, which integrate cell cycle machinery with DNA damage signaling; the response is involved both in DNA repair and apoptosis (Shiloh, 2003; Lukas et al., 2004). Traditionally, neurons have been considered to be "locked" into the G0 phase of the cell cycle. Normally, the release of a cell from the resting G0 phase results in its entry into the first gap phase (G1), during which the cell prepares for DNA replication in the S phase. This is followed by the second gap phase (G2) and mitosis (M phase). While a number of studies have investigated the responses of proliferating cells to genotoxic agents, little is known about the DNA damage response in neurons and other terminally differentiated cells. Differentiated neurons are characterized by the reduction in the cellular levels of several DNA damage response proteins (Lukas et al., 2004; Nouspikel and Hanawalt, 2002; McMurray, 2005; Biton et al., 2006). While this observation is in keeping with the notion that resting cells must reenter the cell cycle in order to carry out DNA repair, a role for cell cycle activation in terminally differentiated cells remains unclear. This activation may be involved in the DNA damage response, DNA repair and apoptosis (Kruman, 2004; Kruman et al., 2004a). Evidence is emerging that in neurodegenerative diseases, neurons attempt to re-enter the cell cycle and this attempt is followed by programmed cell death (Kruman, 2004; Herrup et al., 2004). However, while terminally differentiated neurons retain the ability to reactivate the cell cycle, it rarely leads to neuronal proliferation but typically induces apoptosis (Wartiovaara et al., 2002; Husseman et al., 2000 and Liu and Greene, 2001). Consistent with this idea are the rare incidence of brain tumors of neuronal origin and the resistance of neurons to oncogenic transformation (Heintz, 1993). On the other hand, aneuploidy was demonstrated to be quite common in the brain of adult animals (Rehen et al., 2001), meaning that DNA replication in terminally differentiated cells may lead to aneuploidy. Nevertheless, re-expression and activation of cell cycle proteins have been observed in dying neurons of brains of patients with neurodegenerative disorders (Nagy et al., 1997; Husseman et al., 2000; Yang et al., 2001). Recently, a direct interaction between cell cycle machinery and the cell-intrinsic apoptotic program was demonstrated in postmitotic neurons with the observation that cdc2, a cell cycle regulator, directly initiates apoptosis via direct activation of Bad, a trigger of apoptosis (Konishi et al., 2002). The suppression of cell cycle-dependent kinase (CDK) activity has been shown to produce a neuroprotective effect both *in vivo* and *in vitro* (Park et al., 1997; O'Hare et al., 2002; Katchanov et al., 2001; Rideout et al., 2003), supporting the idea that cell cycle reentry underlies neuronal apoptosis. The suppression of ATM, a key player in the

DNA damage response, was found to attenuate the damage-induced attempts to reenter cell cycle and consequently apoptosis following DNA damage in neurons (Kruman et al., 2004a). This can be linked to findings that neurons from ATM-deficient mice have demonstrated a striking resistance to DNA damage-induced apoptosis (Herzog et al., 1998; Chong et al., 2000; Macleod et al., 2003; Kruman et al., 2004a). Importantly, ATM deficiency rescued the apoptosis of neurons observed in *Lig4*, *Ku70* or *Ku80*- knockout mice (Nussenzweig et al., 1996; Gu et al., 1997; Lee et al., 1997; Sekiguchi et al., 2001). This suggests that the disturbance of DSB repair resulting from the lack of NHEJ components (Lig4, Ku70 or Ku80) in neurons can be recognized by ATM in a pathway leading to apoptosis. In the developing nervous system, this process may serve to remove cells that suffer irreparable damage from the system (McKinnon, 2001), indicating that ATM may function to eliminate neural cells that have incurred DNA damage. On the other hand, the amount of unrepaired DNA after γ-irradiation was greater in cells from ataxia-telangiectasia patients than in cells from normal individuals (Cornforth and Bedford, 1985), underscoring the importance of ATM for DNA repair. Recently, we and others have shown that the DNA damage response in postmitotic neurons committed to apoptosis involves cell cycle- associated events (Klein et al., 2002; Kruman et al, 2004a). Previously, we found that suppression of the ATM function attenuates apoptosis and abrogates cell cycle reentry in postmitotic cortical neurons exposed to DNA-damaging agents (Kruman et al., 2004a). We have now extended this work by showing that ATM deficiency also affects cell cycle activation associated with DNA repair in neurons (Figure 4), in keeping with the documented reduction in DNA repair efficiency in cells from ataxia-telangiectasia patients (Cornforth and Bedford, 1985). This reflects the intimate relationship between ATM, cell cycle activation, and DNA repair in postmitotic neurons, and further suggests that cell cycle reentry is a requisite component of the response to DNA damage essential for the execution of both apoptosis and DNA repair in neurons. Although it is widely accepted that postmitotic cells and tissues have decreased DNA repair capacity, several findings indicate that postmitotic cells such as muscle cells and neurons may up-regulate the expression of DNA glycosylases in certain conditions (Karahalil et al., 2002). Thus, oxoguanine-DNA glycosylase (OGG1; which recognizes and removes 8-oxoguanine) activity in rat heart increases significantly with age (Souza-Pinto et al., 1999). Up-regulation of mRNA levels of APE/Ref1, a multifunctional enzyme that has both an AP endonuclease activity and is essential for glycosylase initiated BER (Flaherty et al., 2001) in neurons, were shown to be high in certain hypothalamic nuclei, as well as in the hippocampus and cerebellum (Wilson et al., 1996). In brain, up-regulation of BER activity was detected after ischemia-reperfusion injury (Lin et al., 2000). Oxidative stress and oxidative DNA damage induce up-regulation of OGG1 activity associated with this type of DNA damage. In support of these observations, we found up-regulated levels of UNG mRNA in brains of cystathionine-β-synthase-knockout (CBS) mice compared to wild-type mice. CBS mice are characterized by hyperhomocysteinemia (Watanabe et al., 1995), which is associated with one-carbon metabolism impairment (itself associated with uracil misincorporation) (Duthie, 1999). Like the up-regulation of OGG1 in response to oxidative DNA damage, uracil misincorporation mediated by hyperhomocysteinemia induces up-regulation of UNG mRNA. The UNG activity is in general significantly higher in proliferating as compared with nonproliferating tissues (Aprelikova and Tomilin, 1982), and the expression of UNG has been demonstrated to be under cell cycle regulation (Nahelhus et al., 1997; Walsh et al., 1995). The inducibility of the BER enzymes, OGG1, UNG and APE/Ref1 may indicate that BER in general is essential for maintaining genomic integrity in the brain and that up-regulation of expression and/or activity of BER enzymes in brain tissue may be induced by DNA damage during certain challenges (e.g. hyperhomocysteinemia in CBS knockout mice or oxidative stress) and mediated by cell cycle activation in neurons.

In cycling cells, the DNA damage response is comprised of cell cycle arrest at specific checkpoints, presumably to allow time for the damage to be repaired, or to activate the apoptotic program if the damage is too extensive to be repaired (Rhind and Russell, 2000). Cell cycle regulation is integrated with DNA repair mechanisms and even utilizes some common proteins (Slupphaug et al., 2003). ATM activated by DSBs phosphorylates key proteins that lead to cell cycle checkpoint arrest and/or apoptosis (Shiloh, 2003). In postmitotic, terminally differentiated neurons, signaling through cell cycle components is also essential for the response to DNA damage, but in contrast_to cycling cells, which undergo growth arrest at specific checkpoints, the signaling in neurons is associated with activation of the cell cycle.

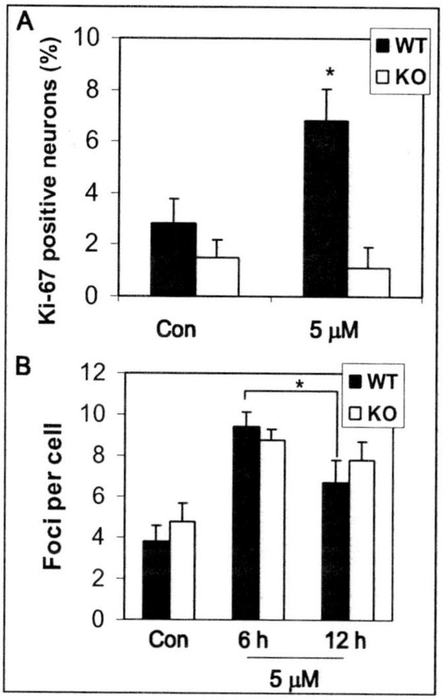


Figure 4. Suppression of ATM function prevents the cell cycle activation in cultured cortical neurons induced by subtoxic (5 μM) concentration of $H_2O_2$, and inhibits DNA repair. (A) Ki-67-positive neurons are reduced in cultures of cortical neurons derived from ATM knockout (ATM-/-) mice compared with those derived from wild-type (WT) following treatment for 12 h with 5 μM $H_2O_2$. (B) Cultures were exposed to either saline or subtoxic (5 μM) concentration of $H_2O_2$, during the indicated periods of time. Immunoreactivity for γ-H2AX was visualized with FITC (488, green). Cells were co-stained with PI. γ-H2AX foci were determined in at least 50 cells. Note significant reduction of γ-H2AX foci in cultures derived from wild-type (WT) mice after 12-h exposure to 5 μM compared to 6-h exposure in contrast to those derived from ATM knockout (ATM-/-) mice. Control (Con) represents the untreated culture. The values are the mean and SD (n=6); $^*p<0.01$.

## Cell Cycle Reentry and Neuronal Apoptosis

Traditionally, neurons have been considered to rest in the G0 phase of the cell cycle. However, accumulating evidence suggests that postmitotic neurons re-express cell cycle markers with the occurrence of neuronal apoptosis (Liu and Greene, 2001; Herrup et al., 2004). These markers include cyclin D, which is involved in the G0-G1 transition, phosphorylated at Ser 795 retinoblastoma tumor suppressor protein (Rb) which is critical for activation of the cell cycle, the DNA helicase subunit minichromosome maintenance (Mcm) proteins 2-7 which assemble in the pre-replication complex, a nuclear antigen Ki-67, which is absent from resting (G0) cells, and cyclin E, a marker of G1-S transition (Becker and Bonni, 2004). Moreover, neurons are able to undergo full or partial DNA replication, showing that they have completed the S phase (Yang et al., 2001; Kruman et al., 2004a). DNA replication has been suggested to be lethal for neurons (Herrup et al., 2004). DNA damage-initiated apoptosis was shown to increase the expression of error-prone DNA polymerase β (Copani et al., 2002), whose physiological role appears restricted to DNA repair (Sobol et al., 1996). In neurons entering the S phase, polymerase β-directed DNA replication might produce additional DNA damage, thus contributing to the execution of apoptosis (Klein et al., 2002).

Little is known about the mechanisms of the DNA damage response in terminally differentiated neurons, especially concerning its integration with cell cycle machinery. However, there is both *in vitro* and *in vivo* evidence of a link between DNA damage and cell cycle reentry in dying postmitotic neurons. Analysis of the X-harlequin (Hq) mutation in the gene encoding apoptosis-inducing factor (AIF) which is accompanied by oxidative stress, has demonstrated that in Hq mice, many cerebellar granule cells had newly synthesized nuclear DNA and were positive for oxo8dG, a marker of oxidative DNA damage (Klein et al., 2002). Using flow cytometry and BrdU incorporation analyses, we have demonstrated that cell cycle activation followed by apoptosis is induced by DNA damage and can be blocked along with the DNA damage response (Kruman et al., 2004a). Although these observations imply that induction of unscheduled cell cycle reentry is highly correlated with, and is likely induced by DNA damage, the mechanisms of the DNA damage response, its link to cell cycle machinery and roles of this machinery in DNA repair of postmitotic neurons are poorly defined. Both the cell cycle and apoptosis are controlled by a highly conserved machinery and exhibit morphological similarities of cell rounding and chromatin condensation. They also share some common participants such as ATM, E2F and p53 (King and Cidlowski, 1995; Matsumura et al., 2003). The *in vivo* evidence that cell cycle re-entry in postmitotic terminally differentiated neurons is associated with DNA damage came from the analysis of the X-linked *harlequin* (*Hq*) mutation in the gene encoding apoptosis-inducing factor (AIF) that causes progressive ataxia (Klein et al., 2002). In *Hq* mice, many but not all cerebellar granule cells had newly synthesized nuclear DNA, as demonstrated by BrdU incorporation, and were positive for caspase-3, indicating the association with apoptosis. In cycling cells, whether DNA damaging conditions cause growth arrest or apoptosis may depend in part on where the cell resides in the cell cycle when the insult is delivered. The fact that not all oxo8dG-positive neurons in brains of *Hq* mice were in the S-phase indicates that these neurons may enter the S-phase later and then die by apoptosis; aqlternately, these oxo8dG-positive neurons may repair the DNA and not enter the S-phase, with subsequent apoptosis, but instead continue normal functioning. Using flow cytometry and BrdU incorporation analyses, we have demonstrated that cell cycle activation followed by apoptosis is induced by DNA damage and can be blocked together with the DNA damage response (Kruman et al., 2004a). These observations and the detection of markers for oxidative DNA damage (long before neurodegeneration) demonstrate that induction of unscheduled cell cycle reentry is highly correlated with, and is likely induced by DNA damage. Furthermore, the amount of oxo8dG, a major oxidative DNA lesion that has been found to be elevated in neuronal DNA in brains of patients with neurodegenerative disorders (Migliore and Coppede, 2002). Increases in oxo8dG are also correlated with increased incidence of cancer (Helbock et al., 1998), endorsing the importance of the same factors in the pathogenesis of both types of disorders.

## Conclusion

In summary, DNA damage may activate cell cycle machinery in postmitotic neurons as an essential part of the DNA damage response. Like in cycling cells, DNA damage may result in DNA repair or apoptosis if DNA damage is unrepaired, and the consequent loss of neurons by apoptosis could result in neurodegeneration. In addition, neurodegeneration is a hallmark of various diseases observed in patients with hereditary DNA repair disorders. The mechanisms by which a cell is forced into apoptosis versus DNA repair, or instead dies by a nonapoptotic mechanism from unrepaired DNA, are not clear. It is likely that both outcomes are managed by the DNA damage response. Suppression of ATM which is a proximal component of DNA damage response, attenuates both apoptosis and cell cycle reentry, suggesting that both cell cycle activation and apoptosis constitute the DNA damage response. ATM deficiency in humans is characterized by predisposition to cancer and progressive neurodegeneration suggesting the impairment of DNA repair when the main player in the DNA damage response does not act. Cortical neurons from ATM-deficient mice are more resistant to apoptosis initiated by DNA damage, suggesting that ATM may function to eliminate neural cells when they have incurred sufficient DNA damage. ATM suppression in neurons prevent cell cycle reentry, while cycling cells deficient in ATM exhibited defective cell cycle checkpoints, implying that neurons may be influenced by the same cell cycle checkpoints that govern apoptosis in cycling cells.

## References

Abraham, RT. (2003). Checkpoint signaling: epigenetic events sound the DNA strand-breaks alarm to the ATM protein kinase. *Bioessays*. *25*,627-630.

Achanta, G; Pelicano, H; Feng, L; Plunkett, W; Huang, P. (2001). Interaction of p53 and DNA-PK in response to nucleoside analogues: potential role as a sensor complex for DNA damage. *Cancer Res 61*,8723-8729.

Alam, ZI; Jenner, A; Daniel, SE; Lees, AJ; Cairns, N;Marsden, CD; Jenner, P; Halliwell, B. (1997). Oxidative DNA damage in the parkinsonian brain: an apparent selective increase in 8-hydroxyguanine levels in substantia nigra. *J Neurochem. 69*,1196-1203.

Aprelikova, O N & Tomilin N V. (1982) Activity of uracil-DNA glycosylase in different rat tissues and in regenerating rat liver. *FEBS Lett. 137*,193-195

Bakkenist, CJ & Kastan, MB. (2003). DNA damage activates ATM through intermolecular autophosphorylation and dimer dissociation. *Nature. 421*,499-506.

Barnes, DE; Stamp, G; Rosewell, I; Denzel, A; Lindahl, T. (1998). Targeted disruption of the gene encoding DNA ligase IV leads to lethality in embryonic mice. *Curr Biol. 8*,1395-1398.

Becker, EB & Bonni, A.(2004). Cell cycle regulation of neuronal apoptosis in development and disease. *Prog Neurobiol. 72*,1-25.

Berkovich, E & Ginsberg, D. (2003). ATM is a target for positive regulation by E2F-1. *Oncogene.22*,161-167.

Bernstein, C; Bernstein, H; Payne CM; Garewal H. (2002). DNA repair/pro-apoptotic dual-role proteins in five major DNA repair pathways: fail-safe protection against carcinogenesis. *Mutat Res 511*,145-178.

Biton, S; Dar, I, Mittelman, L; Pereg, Y; Barzilai, A; Shiloh Y. (2006). Nuclear ataxia-telangiectasia mutated (ATM) mediates the cellular response to DNA double strand breaks in human neuron-like cells. *J Biol Chem. 281*,17482-17491.

Blattner, C; Tobiasch, E; Litfen, M; Rahmsdorf, HJ; Herrlich, P. (1999). DNA damage induced p53 stabilization: no indication for an involvement of p53 phosphorylation. *Oncogene. 18*,1723-1732.

Bogdanov, M; Brown, RH; Matson, W; Smart, R; Hayden, D; O'Donnell; H, Flint Beal; M, Cudkowicz, M. (2000). Increased oxidative damage to DNA in ALS patients. *Free Radic Biol Med. 29*,652-658.

Brooks, PJ. DNA repair in neural cells: basic science and clinical implications (2002). *Mut Res. 509*,93.-108

Boudsocq, F; Ling, H; Yang, W; Woodgate, R. (2002). Structure-based interpretation of missense mutations in Y-family DNA polymerases and their implications for polymerase function and lesion bypass. *DNA Repair (Amst). 1*,343-358.

Burek, MJ & Oppenheim, RW (1996). Programmed cell death in the developing nervous system. *Brain Pathol. 6*,427-446.

Burma, S & Chen, DJ. (2004). Role of DNA-PK in the cellular response to DNA double-strand breaks. *DNA Repair (Amst). 3*,909-918.

Butterfield, DA. (2002). Amyloid beta-peptide (1-42)-induced oxidative stress and neurotoxicity: implications for neurodegeneration in Alzheimer's disease brain. *Free Radic Res. 36*,1307-1313.

Caldecott, KW. (2004). DNA single-strand breaks and neurodegeneration. *DNA Repair (Amst). 3*,875-882.

Carri, NG.(2003). Multiple neurotrophic signalling: certain TGF molecules are involved in retinal development and maturation, but do they complement one another's actions? *Cell Biol Int. 27*,1033-1036.

Chen, J; Jin, K; Chen, M; Pei, W; Kawaguchi, K; Greenberg, DA; Simon, RP. (1997). Early detection of DNA strand breaks in the brain after transient focal ischemia: implications for the role of DNA damage in apoptosis and neuronal cell death. *J Neurochem. 69*,232-245.

Chong, MJ; Murray, MR; Gosink, EC; Russell, HR; Srinivasan, A; Kapsetaki, M; Korsmeyer, SJ; McKinnon, P J. (2000). Atm and Bax cooperate in ionizing radiation-induced apoptosis in the central nervous system. *Proc Natl Acad Sci U S A. 97*.889-894.

Copani, A; Sortino, MA; Caricasole A; Chiechio, S; Chisari, M; Battaglia, G; Giuffrida,-Stella, AM; Vancheri, C; Nicoletti F. (2002). Erratic expression of DNA polymerases by beta-amyloid causes neuronal death. *FASEB J. 16*,2006-2008.

Cornforth; M N & Bedford, JS. (1985). On the nature of a defect in cells from individuals with ataxia-telangiectasia. *Science. 227*,1589-1591.

Cosi, C; Suzuki, H; Skaper, SD; Milani, D; Facci, L; Menegazzi, M; Vantini, G; Kanai, Y; Degryse, A; Colpaert, F; Koek, W; Marien MR. (1997). Poly(ADP-ribose) polymerase (PARP) revisited. A new role for an old enzyme: PARP involvement in neurodegeneration and PARP inhibitors as possible neuroprotective agents. *Ann N Y Acad Sci. 825*,366-379.

Cui, J; Holmes, EH; Greene, TG; Liu, PK. (2000). Oxidative DNA damage precedes DNA fragmentation after experimental stroke in rat brain. *FASEB J. 14*,955-967.

Culmsee, C; Bondada, S; Mattson, MP. (2001). Hippocampal neurons of mice deficient in DNA-dependent protein kinase exhibit increased vulnerability to DNA damage, oxidative stress and excitotoxicity. *Brain Res Mol Brain Res. 87*,257-262.

DeGregori J. (2002). The genetics of the E2F family of transcription factors: shared functions and unique roles. *Biochim Biophys Acta. 1602*,131-150.

Foray, N; Marot, D; Gabriel, A; Randrianarison, V; Carr, AM; Perricaudet, M; Ashworth, A; Jeggo, P. (2003). A subset of ATM- and ATR-dependent phosphorylation events requires the BRCA1 protein. *EMBO J. 22*,2860-28671.

Dimitriadis, EK; Prasad, R; Vaske, MK; Chen, L; Tomkinson, AE; Lewis, MS; Wilson, SH. (1998). Thermodynamics of human DNA ligase I trimerization and association with DNA polymerase beta. *J Biol Chem. 273*,20540-20550.

Duthie, G G. (1999) Determination of activity of antioxidants in human subjects.*Proc. Nutr. Soc. 58*,1015-1024

European Standards Committee on Oxidative DNA Damage (ESCODD). (2002). Comparative analysis of baseline 8-oxo-7,8-dihydroguanine in mammalian cell DNA, by different methods in different laboratories: an approach to consensus. *Carcinogenesis. 23*,2129-2133.

European Standards Committee on Oxidative DNA Damage (ESCODD). (2003). Measurement of DNA oxidation in human cells by chromatographic and enzymic methods. *Free Radic Biol Med. 34*,1089-1099.

Flaherty, DM; Monick, MM; Hunninghake, GW. (2001). AP endonucleases and the many functions of Ref-1. *Am J Respir Cell Mol Biol. 25*,664-667.

Friedberg, EC; McDaniel, LD; Schultz, RA. (2004). The role of endogenous and exogenous DNA damage and mutagenesis. *Curr Opin Genet Dev. 14*,5-10.

Friedlander, RM. (2003). Apoptosis and caspases in neurodegenerative diseases. *N Engl J Med. 348*,1365-1375.

Gao, Y; Sun, Y; Frank, KM; Dikkes, P; Fujiwara, Y; Seidl, KJ; Sekiguchi JM; Rathbun, GA; Swat, W; Wang, J; Bronson, RT; Malynn BA; Bryans, M; Zhu, C; Chaudhuri J; Davidson, L; Ferrini, R J. (1998). *Cell Biol. 119*,493-501.

Gobbel, GT; Bellinzona, M; Vogt, AR; Gupta, N; Fike, JR; Chan, PH. (1998). Response of postmitotic neurons to X-irradiation: implications for the role of DNA damage in neuronal apoptosis. *J Neurosci. 18*,147-55.

Greene, LA; Biswas, SC; Liu, DX. (2004). Cell cycle molecules and vertebrate neuron death: E2F at the hub. *Cell Death Diffe*r. *11*,49-60.

Gu, Y; Seidl, K J ; Rathbun, G A; Zhu, C; Manis, J P; van der Stoep, N; Davidson, L; Cheng, H L; Sekiguchi, J M; Frank, K; Stanhope-Baker, P; Schlissel, MS; Roth, DB; Alt, FW. (1997) Growth retardation and leaky SCID phenotype of Ku70-deficient mice. *Immunity* *7*,653-665.

Harbour, JW & Dean, DC. (2000). The Rb/E2F pathway: expanding roles and emerging paradigms. *Genes Dev.14*,2393-2409.

Heintz, N. (1993). Cell death and the cell cycle: a relationship between transformation and neurodegeneration? *Trends Biochem Sci. 18*,157-159.

Helbock, HJ; Beckman, KB; Shigenaga, MK; Walter, PB; Woodall, AA; Yeo, HC; Ames, BN.(1998). DNA oxidation matters: the HPLC-electrochemical detection assay of 8-oxo-deoxyguanosine and 8-oxo-guanine. *Proc Natl Acad Sci U S A. 95*,288-293.

Herrup, K; Neve, R; Ackerman, SL; Copani, A. (2004). Divide and die: cell cycle events as triggers of nerve cell death. *J Neurosci. 24*,9232-9239.

Herzog, KH; Kapsetaki, M; Morgan, JI; McKinnon, P.J., (1998). Requirement for ATM in ionizing radiation-induced cell death in the developing central nervous system. *Science 280*,1089-1091.

Herzog, KH; Chong, MJ; Kapsetaki, M; Morgan, JI; McKinnon PJ. (1998). Requirement for Atm in ionizing radiation-induced cell death in the developing central nervous system. *Science. 280*,1089-1091.

Ho, L & Hanawalt, PC. (1991).Gene-specific DNA repair in terminally differentiating rat myoblasts. *Mutat Res. 255*,123-141.

Hoeijmakers, JH. (2001). Genome maintenance mechanisms for preventing cancer. *Nature. 411*,366-374.

Honig, LS & Rosenberg, RN.(2000). Apoptosis and neurologic disease. *Am J Med.108*,317-330.

Husseman, JW; Nochlin, D; Vincent, I. (2000). Mitotic activation: A convergent mechanism for a cohort of neurodegenerative disease. *Neurobiol. Aging. 21*,815-828.

Iida, T; Furuta, A; Nishioka, K; Nakabeppu, Y; Iwaki, T. (2002). Expression of 8-oxoguanine DNA glycosylase is reduced and associated with neurofibrillary tangles in Alzheimer's disease brain. *Acta Neuropathol (Berl). 103*,20-25.

Iliakis, G; Wang, Y; Guan, J; Wang, H. (2003). DNA damage checkpoint control in cells exposed to ionizing radiation. *Oncogene. 22*,5834-5847.

Jackson, S. (2002). Sensing and repairing DNA double strand breaks. *Carcinogenesis. 23*, 687-696.

Jeggo, PA & Lobrich M. (2006). Contribution of DNA repair and cell cycle checkpoint arrest to the maintenance of genomic stability. *DNA Repair (Amst)*. [Epub ahead of print]

Jenner, P & Olanow, CW. (1998). Understanding cell death in Parkinson's disease. *Ann Neurol. 44* (Suppl 1),S72-84

Johnson, RT; Gotoh, E; Mullinger, AM; Ryan, AJ; Shiloh, Y; Ziv, Y; Squires, S. (1999). Targeting in double-strand breaks to replicating DNA identifies a subpathway of DSB repair that is defective ataxia-telangiectasia cells. *Biochem Biophys Res Commun. 261*,317-325.

Kannan, K & Jain, SK. (2000). Oxidative stress and apoptosis. *Pathophysiology. 7*,153-163.

Karahalil, B; Hogue, BA; de Souza-Pinto; NC; Bohr, VA. (2002). Base excision repair capacity in mitochondria and nuclei: tissue-specific variations. *FASEB J. 16*,1895-1902.

Karanjawala, ZE; Adachi, N; Irvine, RA; Oh, EK; Shibata, D; Schwarz, K; Hsieh, CL; Lieber, MR. (2002). The embryonic lethality in DNA ligase IV-deficient mice is rescued by deletion of Ku: implications for unifying the heterogeneous phenotypes of NHEJ mutants. *DNA Repair (Amst). 1*, 1017-1026.

Katchanov; J; Harms; C; Gertz; K; Hauck; L; Waeber; C; Hirt; L; Priller; J; von Harsdorf; R; Bruck, W; Hortnagl, H; Dirnagl, U; Bhide, PG; Endres, M. (2001). Mild cerebral ischemia induces loss of cyclin-dependent kinase inhibitors and activation of cell cycle machinery before delayed neuronal cell death. *J Neurosci. 21*,5045-5053.

King, KL & Cidlovski, JA (1995). Cell cycle and apoptosis: common pathways to life and death. *J Cell Biochem. 58*,175-180.

Klein, JA & Ackerman, SL. (2003). Oxidative stress, cell cycle, and neurodegeneration. *J Clin Invest. 111*,785-793.

Klein, JA; Longo-Guess, CM; Rossmann, MP; Seburn, KL; Hurd, RE; Frankel, WN; Bronson, RT; Ackerman, SL. (2002). The harlequin mouse mutation downregulates apoptosis-inducing factor. *Nature. 419*,367-374.

Knudsen, KE; Booth, D; Naderi, S; Sever-Chroneos, Z; Fribourg, AF; Hunton, IC; Feramisco, JR; Wang, JY; Knudsen; ES. (2000). RB-dependent S-phase response to DNA damage. *Mol Cell Biol. 20*,7751-63.

Konishi, Y; Lehtinen, M; Donovan, N; Bonni, A. (2002). Cdc2 phosphorylation of BAD links the cell cycle to the cell death machinery. *Mol Cell. 9*,1005-1016.

Kruman, II (2004). Why do neurons enter the cell cycle? *Cell Cycle. 3*,769-773.

Kruman, II; Wersto, RP; Cardozo-Pelaez, F; Smilenov, L; Chan, SL; Chrest, FJ; Emokpae, R Jr; Gorospe, M; Mattson, MP. 2004a. Cell cycle activation linked to neuronal cell death initiated by DNA damage. *Neuron. 41*,549-561.

Kruman, I; Schwartz, E; Kruman, Y; Cutler, RG; Zhu, X; Greig, NH; Mattson, MP.(2004). Suppression of uracil-DNA glycosylase induces neuronal apoptosis. *J Biol Chem. 279*,43952-60.

Kunkel, TA & Bebenek, K. (2000). DNA replication fidelity. *Annu Rev Biochem. 69*,497-529.

Larsen, NB; Rasmussen, M; Rasmussen, LJ. (2005). Nuclear and mitochondrial DNA repair: similar pathways? *Mitochondrion. 5*,89-108.

Larsen, E; Kwon, K; Coin, F; Egly, JM; Klungland, A. (2004). Transcription activities at 8-oxoG lesions in DNA. *DNA Repair (Amst). 3*,1457-1468.

Lee, SE; Mitchell, RA; Cheng, A; Hendrickson, EA. (1997). Evidence for DNA-PK-dependent and -independent DNA double-strand break repair pathways in mammalian cells as a function of the cell cycle. *Mol Cell Biol. 17*,1425-1433.

Lewen, A; Matz, P; Chan, PH.(2000). Free radical pathways in CNS injury. *J Neurotrauma. 17*,871-890.

Lieber, MR. (1999). The biochemistry and biological significance of nonhomologous DNA end joining: an essential repair process in multicellular eukaryotes. *Genes Cells. 4*, 77-85.

Lin, LH; Cao, S; Yu, L; Cui, J; Hamilton, WJ; Liu, PK.(2000). Up-regulation of base excision repair activity for 8-hydroxy-2'-deoxyguanosine in the mouse brain after forebrain ischemia-reperfusion. *J Neurochem. 74*,1098-10105.

Lin, SC; Skapek, SX; Papermaster, DS; Hankin, M; Lee, EY. (2001). The proliferative and apoptotic activities of E2F1 in the mouse retina. *Oncogene. 20*,7073-7084.

Lips, J & Kaina, B. (2001). DNA double-strand breaks trigger apoptosis in p53-deficient fibroblasts. *Carcinogenesis. 22*,579-585.

Liu, DX & Greene, LA.(2001). Regulation of neuronal survival and death by E2F-dependent gene repression and derepression. *Neuron. 32*,425-438.

Longerich, S & Storb, U. (2005). The contested role of uracil DNA glycosylase in immunoglobulin gene diversification. *Trends Genet. 21*,253-256.

Lovell, MA & Markesbery, WR. (2001). Ratio of 8-hydroxyguanine in intact DNA to free 8-hydroxyguanine is increased in Alzheimer disease ventricular cerebrospinal fluid. *Arch Neurol. 58*,392-396.

Lukas, J; Lukas, C; Bartek, J. (2004). Mammalian cell cycle checkpoints: signalling pathways and their organization in space and time. *DNA Repair (Amst). 3*,997-1007.

Ma, Y; Pannicke, U; Schwarz, K; Lieber, MR. (2002). Hairpin opening and overhang processing by an Artemis/DNA-dependent protein kinase complex in nonhomologous end joining and V(D)J recombination. *Cell. 108*,781-794.

Macleod, MR; Ramage, L; McGregor, A; Seck, JR. (2003). Reduced NMDA-induced apoptosis in neurons lacking ataxia telangiectasia mutated protein. *NeuroReport. 14*,215-217.

MacPhail, S; Banath, J; Yu, T; Chu, E; Lambur, H; Olive, P. (2003). Expression of phosphorylated histone H2AX in cultured cell lines following exposure to X-rays. *Int J Radiat Biol, 79*,351–358.

Mahmoudi, M; Mercer, J; Bennett, M. (2006). DNA damage and repair in atherosclerosis. *Cardiovasc Res. 71*,259-268.

Martin, FL; Williamson, SJ; Paleologou, KE; Hewitt, R; El-Agnaf, OM; Allsop, D. (2003). Fe(II)-induced DNA damage in alpha-synuclein-transfected human dopaminergic BE(2)-M17 neuroblastoma cells: detection by the Comet assay. *J Neurochem. 87*,620-630.

Matsumura, S; Matsumura, T; Ozeki, S,; Fukushima, S; Yamazaki, H; Inoue, T; Inoue, T; Furusawa, Y; Eguchi-Kasai, K.(2003). Comparative analysis of G2 arrest after irradiation with 75 keV carbon-ion beams and 137Cs gamma-rays in a human lymphoblastoid cell line. *Cancer Detect Prev. 27*,222-228.

McKinnon, PJ.(2001). Ataxia telangiectasia: new neurons and ATM. T*rends Mol Med. 7*,233-4.

McMurray, CT.(2005). To die or not to die: DNA repair in neurons. *Mutat Res. 577*,260-274

Mecocci, P; Polidori, MC; Ingegni, T; Cherubini, A; Chionne, F; Cecchetti, R; Senin, U. (1998). Oxidative damage to DNA in lymphocytes from AD patients. *Neurology. 51*,1014-1017.

Meek, DW. (2004). The p53 response to DNA damage. *DNA Repair (Amst). 3*,1049-1056.

Meng, RD; Phillips, P; El-Deiry, WS (1999). p53-independent increase in E2F-1 expression enhances the cytotoxic effects of etoposide and of adriamycin. *Int J Oncol. 14*,5-14.

Migliore, L & Coppede, F. (2002). Genetic and environmental factors in cancer and neurodegenerative diseases. *Mutat Res. 512*,135-153.

Morris, EJ & Geller, HM. (1996). Induction of neuronal apoptosis by camptothecin, an inhibitor of DNA topoisomerase-I: evidence for cell cycle-independent toxicity. *J Cell Biol. 134*, 757-770.

Mullaart, E; Boerrigter, ME; Ravid, R; Swaab, DF; Vijg, J. (1990). Increased levels of DNA breaks in cerebral cortex of Alzheimer's disease patients. *Neurobiol Aging. 11*,169-173.

Nahelhus, TA; Haug, T; Singh, KK; Keshav, KF; Skorpen, F; Otterlei, M; Bharati, S; Lindmo, T; Benichou, S; Benarous, R; Krokan, HE. (1997). A sequence in the N-terminal region of human uracil- DNA glycosylase with homology to XPA interacts with the C-terminal part of the 34-kDa subunit of replication protein A. *J Biol Chem. 272*,6561-6566.

Nagy, Z; Esiri, MM; Cato, AM; Smith, AD.(1997). Cell cycle markers in the hippocampus in Alzheimer's disease. *Acta Neuropathol (Berl). 94*,6-15.

Nazarov, I; Smirnova, A; Krutilina, R; Svetlova, MP; Solovjeva, LV; Nikiforov, AA; Oei, SL; Zalenskaya, IA; Yau, PM.; Bradbury, EM; Tomilin, NV. (2003). Dephosphorylation of histone g-H2AX during repair of DNA double-strand breaks in mammalian cells and its inhibition by calyculin A. *Radiat Res. 160*,309–317.

Niida, H & Nakanishi, M. (2006). DNA damage checkpoints in mammals. *Mutagenesis. 21*,3-9.

Nilsen, H; Otterlei, M; Haug, T; Solum, K; Nagelhus, TA; Skorpen, F; Krokan, HE (1997). Nuclear and mitochondrial uracil-DNA glycosylases are generated by alternative splicing and transcription from different positions in the UNG gene. *Nucleic Acids Res. 25*,750-755.

Nouspikel, T & Hanawalt, PC. (2002). DNA repair in terminally differentiated cells. *DNA Repair (Amst). 1*,59-75.

Nussenzweig, A; Chen, C; da Costa Soares, V; Sanchez, M; Sokol, K; Nussenzweig, MC; Li, GC.(1996) Requirement for Ku80 in growth and immunoglobulin V(D)J recombination. *Nature. 382*,551-555.

O'Connor, DJ & Lu, X. (2000). Stress signals induce transcriptionally inactive E2F-1 independently of p53 and Rb. *Oncogene. 19*,2369-2376.

Offord, E; van Poppel, G; Tyrrell, R. (2000). Markers of oxidative damage and antioxidant protection: current status and relevance to disease. *Free Radic Res. 33* Suppl, S5-19.

O'Hare, M; Wang, F; Park, DS. (2002). Cyclin-dependent kinases as potential targets to improve stroke outcome. *Pharmacol Ther. 93*,135-143.

Oppenheim, RW. (1991). Cell death during development of the nervous system. *Annu Rev Neurosci. 14*,453-501.

Parikh, SS; Mol, CD; Tainer, JA, (1997). Base excision repair enzyme family portrait: integrating the structure and chemistry of an entire DNA repair pathways. *Structure* *5*,1543-1550.

Park, DS; Levine, B; Ferrari, G; Greene, L (1997). Cyclin dependent kinase inhibitors and dominant negative cyclin dependent kinase 4 and 6 promote survival of NGF-deprived sympathetic neurons. *J. Neurosci. 17*,8975-8983.

Park, DS; Morris, EJ; Padmanabhan, J; Shelanski, MI; Geller, HM; Greene LA. (1998). Cyclin-dependent kinases participate in death of neurons evoked by DNA-damaging agents. *J.Cell. Biol. 143*,457-467.

Pfeiffer, P; Goedecke, W; Obe, G. (2000). Mechanisms of DNA double-strand break repair and their potential to induce chromosomal aberrations. *Mutagenesis. 15*,289-302.

Polager, S; Kalma, Y; Berkovich, E; Ginsberg, D. (2002). E2Fs up-regulate expression of genes involved in DNA replication, DNA repair and mitosis. *Oncogene. 21*,437-446.

Przedborski, S & Vila, M. (2003). The 1-methyl-4-phenyl-1,2,3,6-tetrahydropyridine mouse model: a tool to explore the pathogenesis of Parkinson's disease. *Ann N Y Acad Sci. 991*,189-198.

Ratan, RR; Murphy; TH; Baraban, JM. (1994). Oxidative stress induces apoptosis in embryonic cortical neurons. *J Neurochem. 62*,376-379.

Rehen, SK; McConnell, MJ; Kaushal, D; Kingsbury, MA; Yang, AH;Chun, J.(2001). Chromosomal variation in neurons of the developing and adult mammalian nervous system. *Proc Natl Acad Sci U S A. 98*,13361-13366.

Ren, J; Datta, R; Shioya, H; Li, Y; Oki, E; Biedermann, V; Bharti, A; Kufe, D. (2002). p73beta is regulated by protein kinase Cdelta catalytic fragment generated in the apoptotic response to DNA damage. *J Biol Chem. 277*,33758-33765.

Rhind, N & Russell, P. (2000). Checkpoints: it takes more than time to heal some wounds. *Curr Biol. 10*,R908-R911.

Rideout, HJ; Wang, Q; Park, DS; Stefanis, L. (2003). Cyclin-dependent kinase activity is required for apoptotic death but not inclusion formation in cortical neurons after proteasomal inhibition. *J Neurosci 23*,1237-1245.

Rolig, RL & McKinnon, PJ.(2000). Linking DNA damage and neurodegeneration. *Trends Neuroscie. 23*,417- 424.

Rothkamm, K & Lobrich, M. (2003). Evidence for a lack of DNA double-strand break repair in human cells exposed to very low x-ray doses. *Proc Natl Acad Sci USA. 100*:5057–5062.

Sancar A, Lindsey-Boltz LA, Unsal-Kacmaz K, Linn S. (2004). Molecular mechanisms of mammalian DNA repair and the DNA damage checkpoints. Annu Rev Biochem. *73*,39-85.

Sedelnikova, OA; Rogakou, EP; Panyutin, IG; Bonner, WM. (2002). Quantitative detection of (125)IdU-induced DNA double-strand breaks with gamma-H2AX antibody. *Radiat Res. 158*,486-492.

Sedelnikova, OA; Pilch, DR; Redon, C; Bonner, WM. (2003). Histone H2AX in DNA damage and repair. *Cancer Biol Ther. 2*,233-235.

Sekiguchi, J; Ferguson, DO; Chen, HT; Yang, EM; Earle, J; Frank, K; Whitlow, S; Gu, Y; Xu, Y; Nussenzweig, A; Alt, FW. (2001). Genetic interactions between ATM and the nonhomologous end-joining factors in genomic stability and development. *Proc Natl Acad Sci U S A. 98*,3243-3248.

Shi, Y; Venkataraman, SL; Dodson, GE; Mabb, AM; LeBlanc, S; Tibbetts, RS. (2004). Direct regulation of CREB transcriptional activity by ATM in response to genotoxic stress. *Proc Natl Acad Sci U S A. 101*,5898-903.

Shiloh, Y. (2006). The ATM-mediated DNA-damage response: taking shape. *Trends Biochem Sci. 31*,402-410.

Shiloh, Y. (2003). ATM and related protein kinases: safeguarding genome integrity. *Nat Rev Cancer. 3*,155-168.

Slupphaug, G; Kavli, B; Krokan, HE (2003). The interacting pathways for prevention and repair of oxidative DNA damage. *Mutat Res. 531*,231-251.

Sobol, RW; Horton, JK; Kuhn, R; Gu, H; Singhal, RK; Prasad, R; Rajewsky, K, Wilson, SH. (1996). Requirement of mammalian DNA polymerase-beta in base-excision repair. *Nature. 379*, 183-186.

Sonoda, E; Hochegger, H; Saberi, A; Taniguchi, Y; Takeda, S. (2006). Differential usage of non-homologous end-joining and homologous recombination in double strand break repair. *DNA Repair (Amst).* Jun 23; [Epub ahead of print]

Souza-Pinto, NC; Croteau, DL; Hudson, EK; Hansford, RG; Bohr VA.(1999). Age-associated increase in 8-oxo-deoxyguanosine glycosylase/AP lyase activity in rat mitochondria. *Nucleic Acids Res. 27*,1935-1942.

Stevaux, O & Dyson, NJ.(2002). A revised picture of the E2F transcriptional network and RB function. *Curr Opin Cell Biol. 14*,6846-6891.

Stommel, JMn & Wahl, GM. (2004). Accelerated MDM2 auto-degradation induced by DNA-damage kinases is required for p53 activation. *EMBO J. 23*,1547-1556.

Tibbetts, RS; Cortez, D; Brumbaugh, KM; Scully, R; Livingston, D; Elledge, SJ; Abraham, RT. (2000). Functional interactions between BRCA1 and the checkpoint kinase ATR during genotoxic stress. *Genes Dev. 14*,2989-3002.

Tobita, M; Nagano, I; Nakamura, S; Itoyama, Y; Kogure, K. (1995). DNA single-strand breaks in postischemic gerbil brain detected by in situ nick translation procedure. *Neurosci Lett. 200*,129-132.

Veglia, F; Matullo, G; Vineis, P. (2003). Bulky DNA adducts and risk of cancer: a meta-analysis. *Cancer Epidemiol Biomarkers Prev. 12*,157-160.

Vemuri, MC; Schiller, E; Naegele, R. (2001). Elevated DNA double strand breaks and apoptosis in the CNS of scid mutant mice. *Cell Death Differ. 8*,245-255.

Verdaguer, E; Jorda, EG; Stranges, A; Canudas, AM; Jimenez, A; Sureda, FX;, Pallas, M; Camins, A. Inhibition of CDKs: a strategy for preventing kainic acid-induced apoptosis in neurons. *Ann N Y Acad Sci 1010*,671-674.

Walsh, MJ; Shue, G; Spidoni, K; Kapoor, A. (1995). E2F-1 and a cyclin-like DNA repair enzyme, uracil-DNA glycosylase, provide evidence for an autoregulatory mechanism for transcription. *J. Biol. Chem. 270*,5289-5298.

Wang, H; Perrault, AR; Takeda, Y; Qin, W; Wang, H; Iliakis, G. (2003). Biochemical evidence for Ku-independent backup pathways of NHEJ. *Nucleic Acids Res. 3*,5377-5388.

Wartiovaara, K; Barnabe,-Heider, F; Miller, FD; Kaplan, DR. (2002). N-myc promotes survival and induces S-phase entry of postmitotic sympathetic neurons. *J Neurosci. 22*,815-824.

Watanabe, M; Osada, J; Aratani, Y; Kluckman, K; Reddick, R; Malinow, MR; Maeda, N.(1995). Mice deficient in cystathionine beta-synthase: animal models for mild and severe homocyst(e)inemia. *Proc Natl Acad Sci U S A. 92*,1585-1589.

Williams, GM & Jeffrey, AM. (2000). Oxidative DNA damage: endogenous and chemically induced. *Regul Toxicol Pharmacol. 32*,283-292.

Wilson, TM; Rivkees, SA; Deutsch, WA; Kelley, MR.(1996). Differential expression of the apurinic / apyrimidinic endonuclease (APE/ref-1) multifunctional DNA base excision repair gene during fetal development and in adult rat brain and testis. *Mutat Res. 362*,237-248.

Wyllie, AH; Kerr, JF; Currie, AR. (1980). Cell death: the significance of apoptosis. *Int Rev Cytol. 68*,251-306.

Yang, Y; Geldmacher, DS; Herrup, K. (2001). DNA replication precedes neuronal cell death in Alzheimer's disease. *J Neurosci 21*,2661-2668.

Yang, J; Yu, Y; Hamrick, HE; Duerksen-Hughes, PJ. (2003). ATM, ATR and DNA-PK: initiators of the cellular genotoxic stress responses. *Carcinogenesis. 24*,1571-1580.

Yuan, J & Yankner, BA. (2000). Apoptosis in the nervous system. *Nature. 407*,802-809.

In: Neuronal Cell Apoptosis Research
Editor: Relif J. Shandling, pp. 121-137
ISBN 1-60021-456-8


*Chapter 5*

# PORTRAITS OF NEURONAL APOPTOSIS REVEALED BY GENOMIC ANALYSIS

***Sebastiano Cavallaro****
Functional Genomics Center, Institute of Neurological Sciences, Italian National Research Council, 95123 Catania, Italy

## ABSTRACT

Apoptosis is a key physiological response that occurs during development of the nervous system, resulting in the death of nearly half of the embryonic neuronal population, while aberrant apoptotic mechanisms are thought to contribute significantly to many neurological disorders including Alzheimer's disease. Although many experiments in the past have demonstrated the requirement of *de novo* gene expression during neuronal apoptosis, the complete spectrum of genes involved in distinct temporal domains is mostly unknown. To begin a comprehensive survey of the gene-based molecular mechanisms that underlie neuronal apoptosis, we have used the unprecedented experimental opportunities that genome sequences and the development of DNA microarray technology now provide to perform genome-wide expression analysis in different paradigms of neuronal apoptosis. This new approach offers a genomic view of the changes that accompany neuronal apoptosis and yields new insights into the molecular basis underlying it.

* Correspondence concerning this article should be addressed to Sebastiano Cavallaro, M.D., Ph.D., Istituto di Scienze Neurologiche, CNR, Viale Regina Margherita 6, 95123 Catania, Italy. Phone: +39-095-7262411; Fax: +39-095-7122426; E-mail: s.cavallaro@isn.cnr.it.

# INTRODUCTION

Apoptosis, a main form of programmed cell death, is a key physiological response that occurs during development of the nervous system, ensuring the appropriate number of neuronal cells and resulting in the death of nearly half of the embryonic neuronal population [1]. This intrinsic suicide program, by which a cell orchestrates its own destruction, is morphologically distinguished from necrosis, is characterized by nuclear condensation, chromatin margination, plasma membrane blebbing, fragmentation into apoptotic bodies, and internucleosomal cleavage of genomic DNA [2].

Several studies have suggested that neuronal death associated with Alzheimer's disease (AD) is apoptotic in nature [3,4,5,6,7]. There are also data suggesting that in AD brain neurons may spend a prolonged period of time in a "pre-apoptotic" stage prior to switching to the execution stage eventually leading to cell death [8]. During this stage, an abnormal metabolism of amyloid precursor protein and tau, together with an altered functioning of the proteasome system could progressively channel the neurons to an Alzheimer-like phenotype. However, the pathogenetic relationships linking a complex process such as neuronal apoptotic death, with tau pathology and amyloid deposition, as well as proteasomal dysfunction is to date an open problem. Stimuli for apoptosis in AD likely include increased levels of oxidative stress, perturbed calcium homeostasis, accumulation of aggregating amyloid, and reduced energy availability to cells. Each of these factors, which have been documented to be age-related, could be the result of two main causes: (a) deafferentiation of a given neuronal population; (b) lack of neurotrophic supply normally provided by neurotrophins or by other extracellular messengers. For example, if the nerve afferents of cerebellar granule neurons (CGC) are disconnected with a surgical manipulation, these neurons activate a massive program of apoptosis [9]. If, on the other side, potassium concentrations normally employed for culturing the same neurons are reduced from 25 to 5 mM - a condition mimicking *in vivo* neuronal deafferentiation - these neurons activate an internal program of apoptosis [10]. Therefore, elucidation of the apoptotic signaling pathways underlying neurodegeneration could be critical to enhance current therapies for AD.

Many experiments in the past have demonstrated the requirement of *de novo* gene expression during neuronal apoptosis [10,11]. Although previous studies implicated individual genes or genetic pathways during apoptosis, the complete spectrum of genes involved in distinct temporal domains is mostly unknown. To begin a comprehensive survey of the gene-based molecular mechanisms that underlie neuronal apoptosis, we have recently used the unprecedented experimental opportunities that genome sequences and the development of DNA microarray technology now provide to perform genome-scale expression analysis in two different paradigms of neuronal apoptosis [12].

In this chapter I will illustrate how a genomic approach can be useful to dissecting the multigenic program underlying apoptosis. For a more general description of microarray technology, the reader is referred to other reviews [13,14,15].

# 1. DNA Microarray Analysis of Apoptotic Cerebellar Granule Cells

Primary CGC cultures have been extensively utilized to examine the signal transduction mechanisms underlying neuronal apoptosis. In this *in vitro* paradigm, cerebellar granule neurons develop a differentiated phenotype in the presence of serum and elevated potassium concentrations. After removal of serum and lowering of extracellular potassium from 25 to 5 mM, 50% of granule neurons undergo rapid apoptotic cell death within 24 h [10]. This model of apoptotic cell death presumably mimics the naturally occurring death of 20-30% of granule cells, which is important for matching the number of granule cells with Purkinje cells between the third and fifth week postnatally [16], and has its pathological counterpart occurring in an *in vivo* lesion model of deafferentation of the same neuronal population [9]. In this model, mossy fiber axotomy induces a massive granule cells apoptotic death, pointing out to the crucial role of afferent inputs on the survival of these neurons. To identify gene-based molecular mechanisms that underlie neuronal apoptosis, we performed genome-wide expression profiling by microarray analysis of CGCs during the pre-commitment period of apoptosis [12]. By using oligonucleotide microarrays we monitored mRNA expression profiles in CGCs 3 and 6 h after induction of apoptosis by low [$K^+$] (5 mM) and serum deprivation (-KS).

## 1.1. Distinct Temporal Gene Expression Profiles Associated with Apoptosis

Among the 8740 genes interrogated by the microarrays, 423 genes showed significant changes of gene expression. Among these, 69 were down-regulated and 152 up-regulated at 3 h, whereas 93 genes were down-regulated and 204 were up-regulated at 6 h. A complete list of differentially expressed genes is available online at http://web.tiscali.it/sebastiano_cavallaro.

A hierarchical clustering method was used to groups genes differentially expressed after induction of apoptosis on the basis of similarity in their expression patterns. Genes segregating into six major branches of the dendrogram were assigned to six clusters (Figure 1). Clusters 1 to 3 represent those genes that were down-regulated during apoptosis. The majority of these genes were equally down-regulated at 3 and 6 h (n=79, cluster 1) whereas a smaller group were reduced preferentially at 3 h (n=14, cluster 2) or 6 h (n=45, cluster 3). Clusters 4 to 6 differentiate those genes that were up-regulated during apoptosis. The majority of these genes were equally up-regulated at 3 and 6 h (n=167, cluster 4), whereas a smaller group was up-regulated preferentially at 3 h (n=71, cluster 5) or 6 h (n=47, cluster 6).

## 1.2. Transcriptional and Post-Transcriptional Regulatory Mechanisms

DNA microarray technology measure steady-state levels of mRNAs. To observe which of those genes differentially expressed after induction of apoptosis were regulated at the transcriptional or post-transcriptional level, we compared gene expression profiles of CGCs during potassium and serum deprivation in the absence or presence of a transcription inhibitor, Actinomycin-D (ActD). Our analysis demonstrated that most of the genes expression changes observed during apoptosis were under transcriptional control. Treatment with ActD, in fact, reduced the mRNA-expression of most of the genes which were up- (Figure 2, cluster 4, 6, and 7) or down- (Figure 2, cluster 1) regulated during apoptosis. Few of the genes, however, were not affected (Figure 2, cluster 5, and 6) or actually increased (Figure 2, cluster 2, 5 and 7) following ActD treatment, indicating the existence of post-transcriptional regulatory mechanisms.

## 1.3. Functional Clusters of Coregulated Genes and Pathways

Among the 423 genes differentially expressed in CGCs after induction of apoptosis, 295 were assigned to functional categories and subcategories (Figure 3), and to subcellular compartments (Figure 4) based on their translated products.

Some of the genes implicated by microarray analysis have been previously related to apoptosis in CGCs or other cellular systems, whereas others provide a significant number of unique and novel entry points. In many cases, genes with common biological functions or in the same metabolic pathway showed coordinated expression. For space limitation, I will discuss some of the genes implicated by microarray analysis giving more emphasis to functional clusters of coregulated genes and pathways. For a complete description of the genes implicated, the reader is referred to our previous study [12].

### *1.3.1. Signal Transduction*

*Eicosanoid metabolism.* Increased expression of both the GSH-dependent and GSH-independent prostaglandin (PG) D synthases, together with the prostaglandin F2 receptor negative regulator, was observed after apoptosis induction. PGD-synthases are responsible for the conversion of prostaglandin H2 into prostaglandin D2, the major arachidonic metabolite synthesized in mammalian neurons. These observations are in agreement with ability of these enzymes to induce apoptosis [17] and with other studies linking PGs to the regulation of neuronal cell apoptosis [18].

*G Protein-coupled receptors.* The reduced expression of metabotropic glutamate receptor (mGluR) 7 in apoptotic CGCs is consistent with the survival effects exerted by the mGluR receptor agonists [19] and the reduction of another mGluR during apoptosis of CGCs [20]. The down-regulated endothelial differentiation sphingolipid G-protein-coupled receptor 5 (Edg5) has been involved with numerous biological activities, which include stimulation of cell growth and prevention of apoptosis [21].

Increased expression of cannabinoid receptor 1 is consistent with the apoptotic effects which results from its activation in neurons [22], whereas up-regulation of brain natriuretic factor and its receptor 2 is consistent with the ability of this family of peptides to regulate apoptosis [23]. Differential expression of four different olfactory receptors has never been reported and suggests novel functions for these receptors in addition to those previously characterized [24].

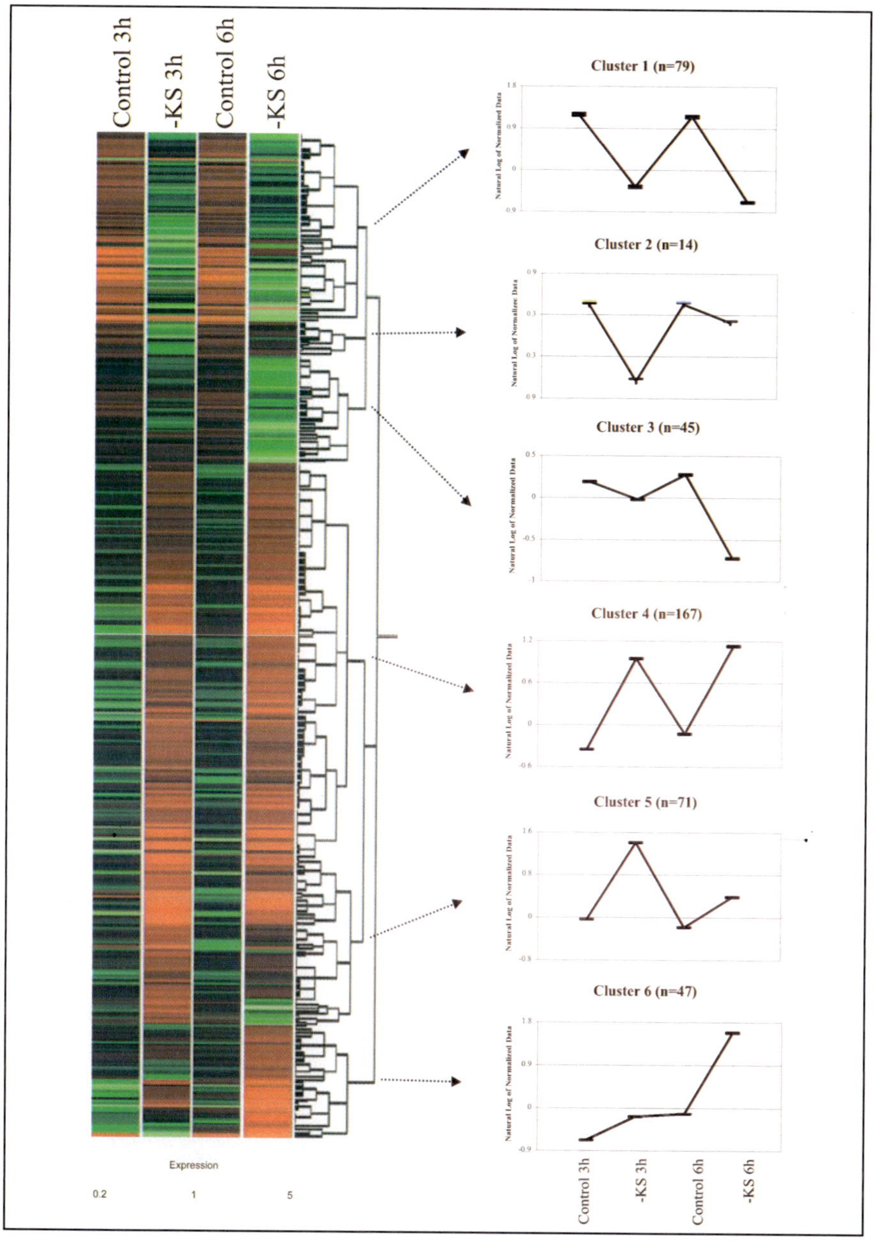


Figure 1. Genes differentially expressed after induction of apoptosis in CGCs. A hierarchical clustering algorithm (Standard correlation, separation ratio 0.5, minimum distance 0.001) was used to order 423 genes differentially expressed 3 and/or 6 h after induction of apoptosis by serum and potassium deprivation. Genes were arranged in a dendrogram in which the pattern and length of the branches reflects the relatedness of the expression levels in four different experimental conditions. Data are presented in a matrix format: each row represents a single gene and each column an experimental condition. The averaged normalized intensity from two replicates is represented by the color of the corresponding cell in the matrix. Green, black and red cells, respectively, represent transcript levels below, equal or above the median abundance across all conditions. Color intensity reflects the magnitude of the deviation from the median (see scale at the bottom). The graphs on the right of the dendrogram represent the averaged Natural Log of normalized data $\pm$ SEM of the genes in six major clusters.

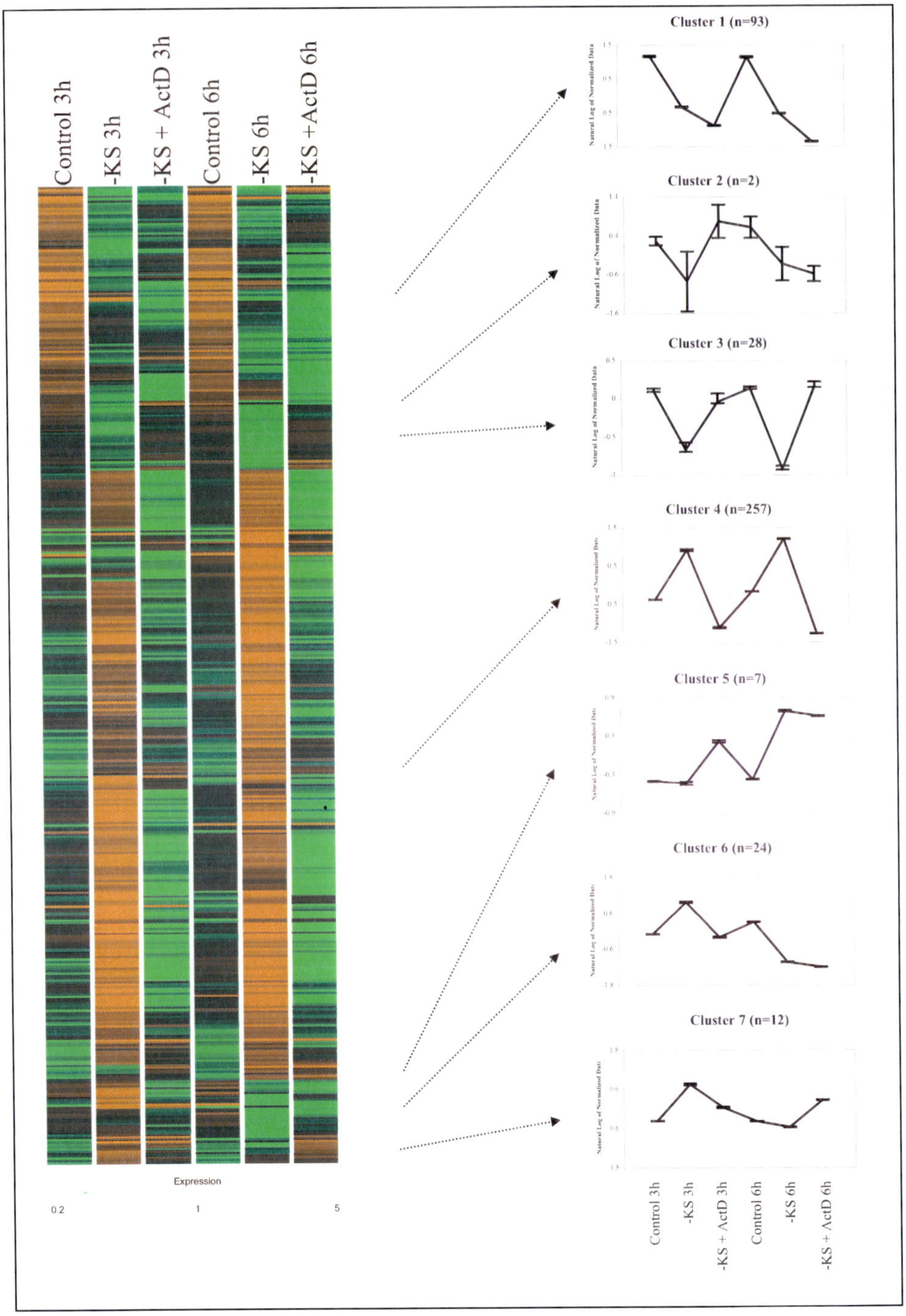


Figure 2. Transcriptional and post-transcriptional regulatory mechanisms during apoptosis in CGCs. A hierarchical clustering algorithm (Pearson correlation, separation ratio 0.5, minimum distance 0.001) was used to order genes differentially expressed after induction of apoptosis in the absence or presence of Actinomycin-D (ActD). Genes were ordered in a dendrogram as specified for Figure 1. The graphs on the right of the dendrogram represent the averaged Natural Log of normalized data ± SEM of the genes in seven major clusters.

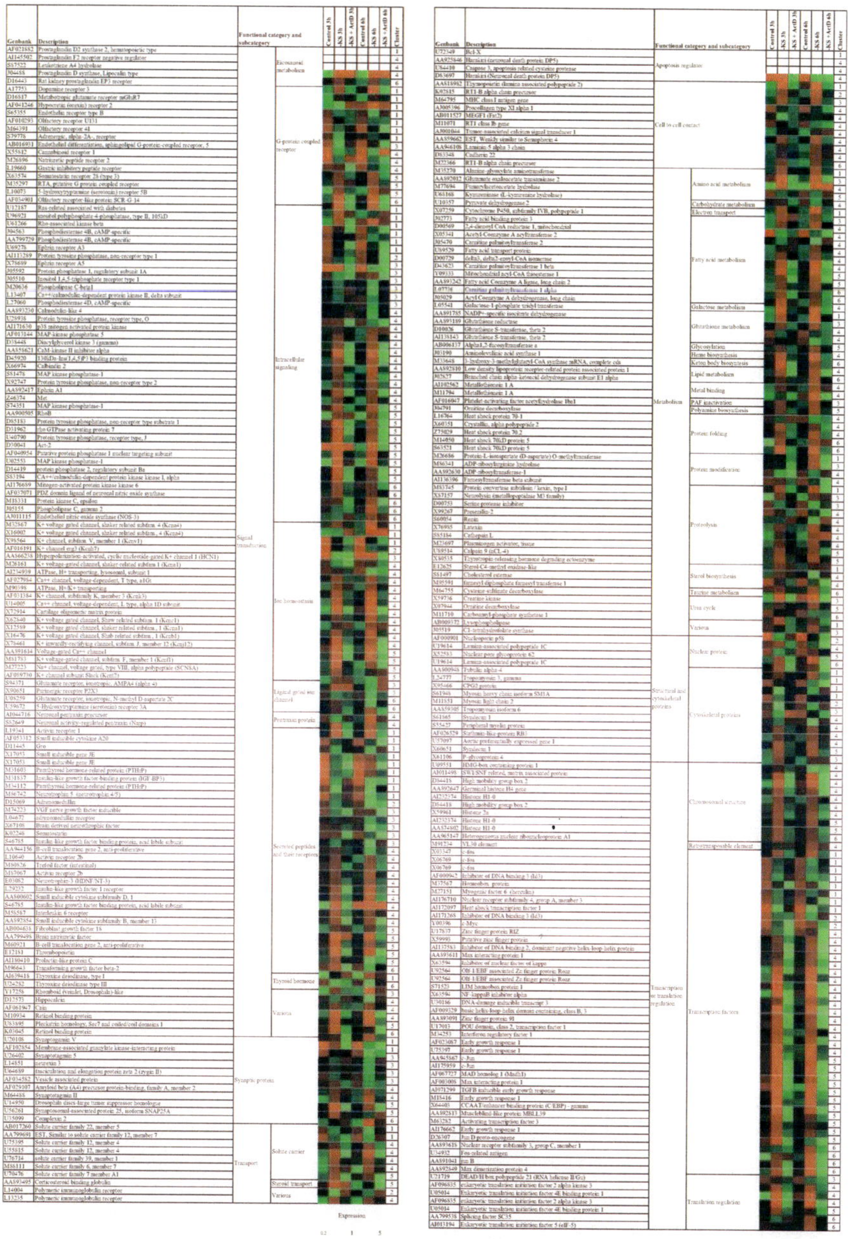

Figure 3. Semantic clustering of genes differentially expressed during apoptosis in CGCs. Genes whose translated products have a recognized function were clustered into different functional categories and subcategories. In order to transform features that are similar functionally in near values, gene ontologies were transformed in numerical values and then clustered by a numerical clustering algorithm (standard correlation). Gene expression level is reported in a matrix format: each row represents a single gene and each column an experimental condition. The averaged normalized intensity from two replicates is represented by the color of the corresponding cell in the matrix as specified for Figure 2. In many cases, genes with common biological functions had similar expression profiles and segregated in the same temporal cluster (the number on the right of the matrix correspond to one of the six major clusters shown in Figure 2). Twenty-eight genes represented by two or three probe sets on the array had similar expression profiles indicating that the effects of experimental noise or artifact were negligible.

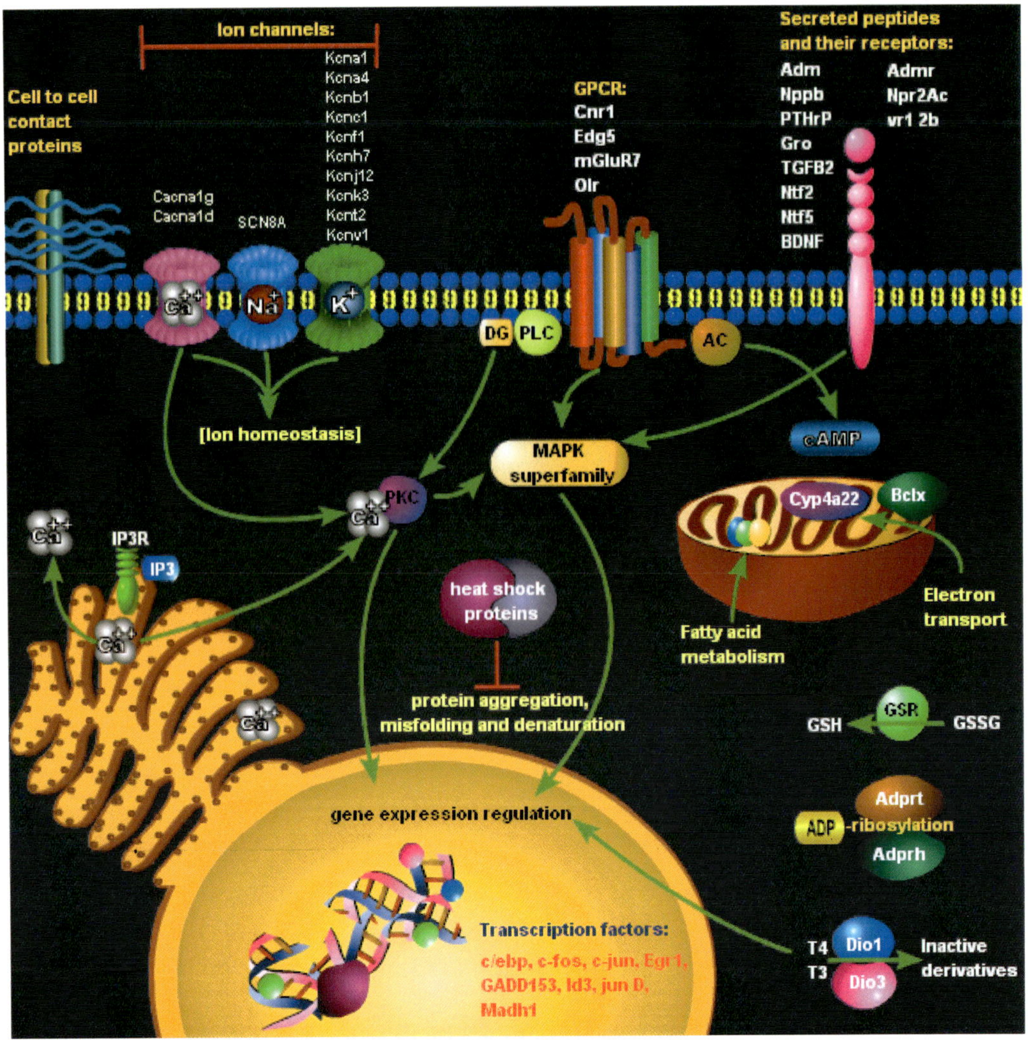


Figure 4. Subcellular localization of proteins encoded by genes differentially expressed after induction of apoptosis in CGCs. Some of the genes whose translated products have a recognized function and subcellular localization are represented into different functional groups and subcellular compartments. *Plasma membrane*. Cell to cell contact proteins. Ion channels: Cacna1g, Ca++ channel, voltage-dependent, T type, a1Gt; Cacna1d, Ca++ channel, voltage-dependent, L type, alpha 1D subunit; SCN8A, Na+ channel, voltage gated, type VIII, alpha polypeptide; Kcna4, K+ voltage gated channel, shaker related subfamily, 4; Kcnv1, K+ channel, subfamily V, member 1; Kcnh7, K+ channel erg3; HCN1, Hyperpolarization-activated, cyclic nucleotide-gated K+ channel 1; Kcna1, K+ voltage-gated channel, shaker-related subfamily, 1; Kcnk3, K+ channel, subfamily K, member 3; Kcnc1, K+ voltage gated channel, Shaw related subfamily, 1; Kcna1, K+ voltage gated channel, shaker related subfamily, 1; Kcnb1, K+ voltage gated channel, Shab related subfamily, 1; Kcnj12, K+ inwardly-rectifying channel, subfamily J, member 12; Kcnf1, K+ voltage-gated channel, subfamily F, member 1; Kcnt2, K+ channel subunit Slack. G-protein coupled receptors (GPCR): Cnr1, cannabinoid receptor 1; Edg5, endothelial differentiation sphingolipid G-protein-coupled receptor 5; mGluR7, metabotropic glutamate receptor; Olr, olfactory receptors (U131, 41, SCR-G-149). Secreted peptides and their receptors: Acvr1 2b, activin receptors 1 and 2b; Adm, adrenomedullin; Admr, adrenomedullin receptor; BDNF, Brain derived Neurothrophic Factor; Gro, Growth Regulated Oncogene; Nppb, brain natriuretic factor Npr2, brain natriuretic factor receptor 2; Ntf2, Neurotrophin 3; Ntf5, Neurotrophin 5; PTHrP, parathyroid hormone-related protein; TGFB2, Transforming Growth Factor Beta-2.

DG, diacylglycerol; PLC, phospholipase-C; AC, adenylate cyclase. *Endoplasmic reticulum***:** IP3, inositol 1,4,5-triphosphate receptor 1. ***Cytosol.*** PKC, protein kinase C gamma. MAPK superfamily (MAPK, p38 mitogen-activated protein kinase; MAPK phosphatases 1 and 5). Heat shock proteins (Heat shock protein 70-1; Crystallin, alpha polypeptide 2; Heat shock protein 70.2; Heat shock 70kD protein 5; Heat shock 70kD protein 5). ADP ribosylation: Adprt, ADP-ribosyltransferase; Adprh, ADP-ribosylarginine hydrolase. ***Mitochondria.*** Electron transport (Cyp4a22, Cytochrome P450, subfamily IVB, polypeptide 1). Fatty acid metabolism (2,4-dienoyl CoA reductase 1; Acetyl-CoA acyltransferase 2; carnitine palmitoyltransferase 1 alpha, 1 beta and 2; delta3, delta2-enoyl-CoA isomerase; acyl-CoA thioesterase 1; fatty acid CoA-ligase; and Acyl CoA dehydrogenase). Bclx, Bcl2-like 1. ***Cytosol.*** PKC, protein kinase C. Glutathione metabolism: GSH, reduced glutathione; GSSG, oxidized glutathione; Gsr, glutathione reductase; Thyroid hormone metabolism: Dio1, deiodinase, iodothyronine, type I; Dio3, deiodinase, iodothyronine, type III; T4, thyroxine; T3, Triiodothyronine. ***Nucleus.*** Transcription factors: c/ebp, CCAAT/enhancer binding protein (C/EBP), gamma; c-fos, FBJ murine osteosarcoma viral oncogene homolog; c-jun, Jun oncogene; Egr1, Early growth response 1; GADD153, growth arrest and DNA damage-inducible gene; Id3, Inhibitor of DNA binding 3; jun D, Jun D proto-oncogene; Madh1, SMAD, mothers against DPP homolog 1.

*Intracellular signaling.* Decreased expression of inositol 1,4,5-triphosphate receptor type 1 has been previously reported during apoptosis [25]. Similarly, the p38 mitogen-activated protein kinase (MAPK) and two MAPK phosphatases 1 and 5, which dephosphorylate both serine/threonine and tyrosine residues and inactivate members of the MAPK superfamily (MAPK/ERK, SAPK/JNK, p38) have been previously associated with apoptotic cell death, cellular proliferation and differentiation [26].

Increased expression of endothelial nitric oxide synthase (NOS)-3 was observed 6 h after apoptotic induction in CGCs and is consistent with the increased expression of NOS after growth factor deprivation and the regulatory effects by its product, nitric oxide, during apoptosis [27]. Up-regulation of RhoB, a member of the family of GTPases that regulate a variety of signal transduction pathways, is in agreement with recent data implicating RhoB in apoptosis [28].

Differential expression of activin receptor 1and 2b is consistent with the known ability of activins, which are members of the transforming growth factor-beta family, to induce autocrine inhibition of cell proliferation and enhancement of apoptosis [29].

*Ion homeostasis.* In line with other reports [30], we observed the differential expression of a number of proteins involved with ion transport. Among these are three voltage dependent $Ca^{2+}$ channels and ten $K^+$ channel subunits (Figure 3 and 4). With the exception of Kcna4, Kcnh7 and Kcnv1, $K^+$ channel subunits were up-regulated in apoptotic CGCs. Up-modulation of $K^+$ channels during apoptosis has been observed in a number of cells and supports the general idea that enhancement of $K^+$ currents mediates apoptosis whereas its blockade prevents it [30]. Potassium efflux may be a step for propagating apoptotic signaling since decreased $[K^+]_i$ is known to regulate caspase activation, cytochrome c release, activation of the JNK/SAPK signaling pathway, chromatin condensation and DNA fragmentation [30].

*Ligand gated ion channels.* Differential expression of two ionotropic glutamate receptors, AMPA4 and NMDA2C, is consistent with previous experiments showing that granule cell survival, both in vivo and in vitro, can be modulated by glutamate receptors [31,32].

*Pentraxin proteins.* Increased expression of two members of the pentraxin protein family, neuronal pentraxin I (NPXI) and neuronal activity–regulated pentraxin (Narp, also known as pentraxin 2), was observed in CGCs undergoing apoptosis. Over expression of NPX1 in apoptotic CGCs has been recently confirmed in another study [33]. Narp was originally isolated as an immediate early gene induced by seizure in the rat hippocampus [34] and then found to have a key role in inducing synaptic clustering of glutamate receptors [35]. The genetic switch to sensor changes in synaptic activity mediated by Narp may be relevant since activation of receptors by receptor clustering can lead to programmed cell death [36].

*Secreted peptides and their receptors.* Decreased expression of parathyroid hormone-related protein (PTHrP) in apoptotic CGCs confirms previous findings showing that its expression is regulated in an activity-dependent manner and it is required for survival by functioning in an autocrine/paracrine loop [37]. Reduced expression of adrenomedullin and its receptor is consistent with the proliferative and anti-apoptotic effects exerted by this peptide [38].

The small inducible cytokine A20, B13, D1 and Growth Regulated Oncogene (Gro, also known as KC), are members of the CXC family of chemokines. Members of this family and their receptors are widely expressed throughout the central nervous system and have been involved in neuronal-programmed cell death. For instance, a protective effect of IL-8 or Gro-beta has been described in cultured hippocampal neurons [39] and CGCs [32].

Increased expression of Transforming Growth Factor Beta-2 (TGFB2) is consistent with its pro-apoptotic effect and its increased syntesis during CGC apoptosis [40]. Differential expression of Neurotrophin (NT) 3 and 5, and Brain derived Neurothrophic Factor (BDNF), were previously reported [41].

### *1.3.2. Thyroid Hormone*

Thyroid hormone (TH) is essential for normal development and its deficiency may result in cretinism and alter the development of the cerebellum producing apoptosis of cerebellar granule cells [42]. Intracellular concentrations of the thyroid hormones, T4 and T3, are influenced by the activity of three thyroxine deiodinases (TD), classified as type I, II, and III. TD-I and -II catalyze the intracellular conversion of thyroid hormone T4 to the biologically active hormone T3, whereas TD-III catalyzes the degradation of both T4 and T3 to inactive derivatives. The observed reduction of TD-I and increase of TD-III in apoptotic CGCs may facilitate apoptosis by influencing the intracellular rates of T4 to T3 conversion.

### *1.3.3. Apoptosis Regulators*

Among those genes that are conventionally designated as apoptosis regulators, we observed the differential expression of Bcl-X, caspase 3, calpain and Harakiri (also known as neuronal death protein DP5). The induction of this last and less characterized gene, which belongs to the family of pro-apoptotic BH3-only proteins, is consistent with its known ability to induce apoptosis in nerve tissue [43].

### *1.3.4. Cell to Cell Contact*

Most of the genes implicated with cell-to-cell contact were found increased during apoptosis of CGCs. One of them, the up-regulated MEGF1/FAT2, encodes a tumor suppressor essential for controlling cell proliferation and playing an important role in cerebellum development [44]. Down-regulation of major histocompatibility complex MHC molecules, such as RT1-B alpha and 1b, has been found in other cells during apoptosis [45].

### 1.3.5. Metabolism

*Fatty acid metabolism.* The coordinated increase of eight enzymes (2,4-dienoyl CoA reductase 1; Acetyl-CoA acyltransferase 2; carnitine palmitoyltransferase 1 alpha, 1 beta and 2; delta3, delta2-enoyl-CoA isomerase; acyl-CoA thioesterase 1; fatty acid CoA-ligase; and Acyl CoA dehydrogenase), which are key regulators of fatty acid metabolism (Figure 3 and 4), were found in apoptotic CGCs. Five of these enzymes, in particular, are involved in the beta-oxidation pathway whose induction may reflect an increased demand in energy production during the energy-requiring apoptotic program, also in view of the rapid and progressive impairment of oxidative phosphorylation during the same time period [46,47].

*Protein folding.* All genes involved with protein folding were up-regulated during apoptosis of CGCs. Most of them are heat shock proteins (HSPs), which act as molecular chaperones to assist other proteins in their folding, transport and assembly into complexes. Although elevated expression of HSPs has been reported during apoptosis [48], it is unclear if HSPs mediate an adaptive response or play a role in the progression of apoptosis.

*Protein modification.* Up-regulation of ADP-ribosyltransferase and the down-regulation of ADP-ribosylarginine hydrolase in apoptotic CGCs is in support of the recent involvement of ADP-ribosylation in apoptosis [49]. ADP-ribosyltransferase, in fact, catalyzes the transfer of ADP-ribose to an arginine residue in proteins, whereas ADP-ribosylarginine hydrolase catalyzes the reverse reaction by removing ADP-ribose and regenerating free arginine.

Protein-L-isoaspartate (D-aspartate) O-methyltransferase is an enzyme which methylates aspartate residues of proteins damaged by age-related isomerisation and deamidation. Down-regulation of this enzyme in CGCs undergoing apoptosis is in line with the recent demonstration of its protective effect from Bax-induced apoptosis [50].

*PAF inactivation.* Reduced expression of platelet-activating factor acetylhydrolase during apoptosis of CGCs is in agreement with the anti apoptotic effect exerted by this enzyme [51].

*Amino acid metabolism.* Increased expression of Kynurinase, an enzyme involved in the "kynurenine pathway" of tryptophan metabolism may lead to accumulation of neurotoxic tryptophan metabolites known to induce apoptosis [52].

*Glutathione metabolism.* Altered expression of glutathione S-transferase and glutathione reductase, together with $NADP^+$-specific isocitrate dehydrogenase a major producer of mitochondrial NADPH that is necessary for the regeneration of GSH, support the role of oxidative stress and reactive oxygen species in apoptosis as also previously surmised following a study of the same neurons under the same apoptotic conditions [47].

*Proteolysis.* Increased expression of presenlin 2, a gene mutated in some cases of familial AD, has been shown to increase the sensitivity to apoptotic stimuli [53,54]. This finding could be interpreted on the light of previous studies demonstrating that following the apoptotic trigger, activation of an amyloidogenic route occurs in CGCs and leads to overproduction of b-amyloid [55]. These events could be the consequence, or at least be part of the pathway, involving an increased expression of proteins such as presenilin 2, directly involved in the amyloidogenic route activated in CGCs during apoptosis.

### 1.3.6. Transcription or Translation Regulation

*Chromosomal structure.* Differential expression of H1-0, H2b and H4 histones, three proteins involved in the formation of higher order chromatin structure, is consistent with their role during apoptotic DNA fragmentation. Increased expression of the high mobility group box (HMGB) 1 and 2 genes, which encode DNA-bending/looping chromosomal proteins, may regulate DNA fragmentation during apoptosis through changes in chromosome structure [56].

*Transcription factors.* Among these are a group of immediate early genes, c-fos, c-jun, jun D, c/ebp gamma and the early growth response 1 (Egr1, also termed NGFI-A, Krox24, or zif/268), which are thought to act as third messengers in the coupling of trans-synaptic stimuli to neuronal gene expression. Decreased expression of c-fos, and up-regulation of c-jun and Egr1 during CGCs apoptosis were previously demonstrated [57,58,59]. Other transcription factors, such as Id3 [60], activating transcription factor 3 [61], GADD153 [62], and Madh1 [63] were also previously implicated during apoptotic cell death.

*Translation regulation.* During CGCs apoptosis we observed the increased expression of eukaryotic translation initiation factor 2 alpha kinase 3 which regulates the frequency of polypeptide chain initiation and is known to modulate apoptosis [64]. This data is in support with the notion that apoptosis in a variety of cell types leads to inhibition of protein synthesis.

## CONCLUSION

While demonstrating the utility of a cDNA-microarray system as a means of dissecting the multigenic program underlying apoptosis, the data presented represent just a glimpse of this complex phenomenon. It should be emphasized that the microarray provides estimates of changes in mRNA levels that cannot be correlated with the amount and function of the gene products. Translation and post-translational modifications of many gene products and protein turnover have dramatic effects on function, and these cannot be inferred from expression analysis alone.

Most of the changes observed in CGCs seem to be pro-apoptotic, while others could be adaptive and represent an attempt for survival. Each of them, having a larger or smaller specific weight may contribute to the development of life or death. The exact role and functional relationships of some the genes implicated cannot yet be recognized. Gene expression profiles unlock virtually unexplored frontiers and we will learn as we explore them. Systematic characterization of expression patterns associated with apoptosis in different pathophysiological conditions and in distinct temporal domains will provide a framework for interpreting the biological significance of the expression patterns observed in the present study. We have just begun such a challenging task by characterizing gene expression profiles in cortical neurons exposed to amyloid β-protein (β–AP), whose toxicity is considered the leading mechanism proposed for neuronal death in AD. By using oligonucleotide microarrays we monitored mRNA expression profiles in primary cultures of rat cortical neurons exposed for 24h or 48h to vehicle, 20 μM β–AP 1-40, and 20 μM β–AP 40-1. Among the genes differentially expressed following βAP 1-40 treatment, 70 were in common with those differentially expressed during apoptosis of CGCs [12]. Although preliminary, these data suggest the existence of both common and diverse mechanisms responsible of neuronal cell death. Knowledge of the mechanisms and pathways that determine apoptosis and are aberrant in pathological conditions will pave the way to new and effective therapeutic approaches.

## Acknowledgements

We gratefully acknowledge Alfia Corsino, Maria Patrizia D'Angelo and Francesco Marino for their administrative and technical support. This work was partly sponsored by grants of the Italian Ministry of Education University and Research (DM 1105/255; FAR 10036) to SC.

## References

[1] Pettmann, B and Henderson, CE. Neuronal cell death. *Neuron,* 1998 20, 633-647.

[2] Arends, MJ and Wyllie, AH. Apoptosis: mechanisms and roles in pathology. *Int. Rev. Exp. Pathol.,* 1991 32, 223-254.

[3] Bredesen, DE. Neural apoptosis. *Ann. Neurol.,* 1995 38, 839-851.

[4] Smale, G; Nichols, NR; Brady, DR; Finch, CE; Horton, WE, Jr. Evidence for apoptotic cell death in Alzheimer's disease. *Exp. Neurol.,* 1995 133, 225-230.

[5] Vincent, I; Jicha, G; Rosado, M; Dickson, DW. Aberrant expression of mitotic cdc2/cyclin B1 kinase in degenerating neurons of Alzheimer's disease brain. *J. Neurosci.,* 1997 17, 3588-3598.

[6] Busser, J; Geldmacher, DS; Herrup, K. Ectopic cell cycle proteins predict the sites of neuronal cell death in Alzheimer's disease brain. *J. Neurosci.,* 1998 18, 2801-2807.

[7] Jellinger, KA and Stadelmann, C. Problems of cell death in neurodegeneration and Alzheimer's Disease. *J. Alzheimers. Dis.,* 2001 3, 31-40.

[8] Yang, Y; Geldmacher, DS; Herrup, K. DNA replication precedes neuronal cell death in Alzheimer's disease. *J. Neurosci.,* 2001 21, 2661-2668.

[9] Borsello, T; Di Luzio, A; Ciotti, MT; Calissano, P; Galli, C. Granule neuron DNA damage following deafferentation in adult rats cerebellar cortex: a lesion model. *Neuroscience,* 2000 95, 163-171.

[10] D'Mello, SR; Galli, C; Ciotti, T; Calissano, P. Induction of apoptosis in cerebellar granule neurons by low potassium: inhibition of death by insulin-like growth factor I and cAMP. *Proc. Natl. Acad. Sci. U. S. A,* 1993 90, 10989-10993.

[11] Galli, C et al. Apoptosis in cerebellar granule cells is blocked by high KCl, forskolin, and IGF-1 through distinct mechanisms of action: the involvement of intracellular calcium and RNA synthesis. *J. Neurosci.,* 1995 15, 1172-1179.

[12] Cavallaro, S et al. Gene expression profiles of apoptotic neurons. *Genomics,* 2004 84, 485-496.

[13] Heller, MJ. DNA Microarray Technology: Devices, Systems, and Applications. *Annu. Rev. Biomed. Eng,* 2002 4, 129-153.

[14] Hess, KR; Zhang, W; Baggerly, KA; Stivers, DN; Coombes, KR. Microarrays: handling the deluge of data and extracting reliable information. *Trends Biotechnol.,* 2001 19, 463-468.

[15] Noordewier, MO and Warren, PV. Gene expression microarrays and the integration of biological knowledge. *Trends Biotechnol.,* 2001 19, 412-415.

[16] Williams, RW and Herrup, K. The control of neuron number. *Annu. Rev. Neurosci.,* 1988 11, 423-453.

[17] Ragolia, L; Palaia, T; Frese, L; Fishbane, S; Maesaka, JK. Prostaglandin D2 synthase induces apoptosis in PC12 neuronal cells. *Neuroreport,* 2001 12, 2623-2628.

[18] Kondo, M et al. 15-Deoxy-Delta(12,14)-prostaglandin J(2): the endogenous electrophile that induces neuronal apoptosis. *Proc. Natl. Acad. Sci. U. S. A,* 2002 99, 7367-7372.

[19] Copani, A et al. Activation of metabotropic glutamate receptors protects cultured neurons against apoptosis induced by beta-amyloid peptide. *Mol. Pharmacol.,* 1995 47, 890-897.

[20] Borodezt, K and D'Mello, SR. Decreased expression of the metabotropic glutamate receptor-4 gene is associated with neuronal apoptosis. *J. Neurosci. Res.,* 1998 53, 531-541.

[21] Takuwa, Y; Takuwa, N; Sugimoto, N. The edg family g protein-coupled receptors for lysophospholipids: their signaling properties and biological activities. *J. Biochem. (Tokyo),* 2002 131, 767-771.

[22] Downer, E; Boland, B; Fogarty, M; Campbell, V. Delta 9-tetrahydrocannabinol induces the apoptotic pathway in cultured cortical neurones via activation of the CB1 receptor. *Neuroreport,* 2001 12, 3973-3978.

[23] Wu, CF; Bishopric, NH; Pratt, RE. Atrial natriuretic peptide induces apoptosis in neonatal rat cardiac myocytes. *J. Biol. Chem.,* 1997 272, 14860-14866.

[24] Glusman, G et al. The olfactory receptor gene superfamily: data mining, classification, and nomenclature. *Mamm. Genome,* 2000 11, 1016-1023.

[25] Oberdorf, J; Vallano, ML; Wojcikiewicz, RJ. Expression and regulation of types I and II inositol 1,4,5- trisphosphate receptors in rat cerebellar granule cell preparations. *J. Neurochem.,* 1997 69, 1897-1903.

[26] Chang, L and Karin, M. Mammalian MAP kinase signalling cascades. *Nature,* 2001 410, 37-40.

[27] Kim, PK; Zamora, R; Petrosko, P; Billiar, TR. The regulatory role of nitric oxide in apoptosis. *Int. Immunopharmacol.,* 2001 1, 1421-1441.

[28] Prendergast, GC. Actin' up: RhoB in cancer and apoptosis. *Nat. Rev. Cancer,* 2001 1, 162-168.

[29] Chen, YG; Lui, HM; Lin, SL; Lee, JM; Ying, SY. Regulation of cell proliferation, apoptosis, and carcinogenesis by activin. *Exp. Biol. Med. (Maywood. ),* 2002 227, 75-87.

[30] Yu, SP; Canzoniero, LM; Choi, DW. Ion homeostasis and apoptosis. *Curr. Opin. Cell Biol.,* 2001 13, 405-411.

[31] Lin, WW; Wang, CW; Chuang, DM. Effects of depolarization and NMDA antagonists on the role survival of cerebellar granule cells: a pivotal role for protein kinase C isoforms. *J. Neurochem.,* 1997 68, 2577-2586.

[32] Limatola, C et al. The chemokine growth-related gene product beta protects rat cerebellar granule cells from apoptotic cell death through alpha-amino-3-hydroxy-5-methyl-4-isoxazolepropionate receptors. *Proc. Natl. Acad. Sci. U. S. A,* 2000 97, 6197-6201.

[33] DeGregorio-Rocasolano, N; Gasull, T; Trullas, R. Overexpression of neuronal pentraxin 1 is involved in neuronal death evoked by low K(+) in cerebellar granule cells. *J. Biol. Chem.,* 2001 276, 796-803.

[34] Tsui, CC et al. Narp, a novel member of the pentraxin family, promotes neurite outgrowth and is dynamically regulated by neuronal activity. *J. Neurosci.,* 1996 16, 2463-2478.

[35] O'Brien, RJ et al. Synaptic clustering of AMPA receptors by the extracellular immediate- early gene product Narp. *Neuron,* 1999 23, 309-323.

[36] Cribbs, DH; Kreng, VM; Anderson, AJ; Cotman, CW. Cross-linking of concanavalin A receptors on cortical neurons induces programmed cell death. *Neuroscience,* 1996 75, 173-185.

[37] Ono, T et al. Activity-dependent expression of parathyroid hormone-related protein (PTHrP) in rat cerebellar granule neurons. Requirement of PTHrP for the activity-dependent survival of granule neurons. *J. Biol. Chem.,* 1997 272, 14404-14411.

[38] Cuttitta, F et al. Adrenomedullin functions as an important tumor survival factor in human carcinogenesis. *Microsc. Res. Tech.,* 2002 57, 110-119.

[39] Araujo, DM and Cotman, CW. Trophic effects of interleukin-4, -7 and -8 on hippocampal neuronal cultures: potential involvement of glial-derived factors. *Brain Res.,* 1993 600, 49-55.

[40] de Luca, A; Weller, M; Fontana, A. TGF-beta-induced apoptosis of cerebellar granule neurons is prevented by depolarization. *J. Neurosci.,* 1996 16, 4174-4185.

[41] Condorelli, DF; Dell'Albani, P; Timmusk, T; Mudo, G; Belluardo, N. Differential regulation of BDNF and NT-3 mRNA levels in primary cultures of rat cerebellar neurons. *Neurochem. Int.,* 1998 32, 87-91.

[42] Xiao, Q and Nikodem, VM. Apoptosis in the developing cerebellum of the thyroid hormone deficient rat. *Front Biosci.,* 1998 3, A52-A57.

[43] Inohara, N; Ding, L; Chen, S; Nunez, G. harakiri, a novel regulator of cell death, encodes a protein that activates apoptosis and interacts selectively with survival-promoting proteins Bcl-2 and Bcl-X(L). *EMBO J.,* 1997 16, 1686-1694.

[44] Dunne, J et al. Molecular cloning and tissue expression of FAT, the human homologue of the Drosophila fat gene that is located on chromosome 4q34-q35 and encodes a putative adhesion molecule. *Genomics,* 1995 30, 207-223.

[45] Fajka-Boja, R et al. Fermented wheat germ extract induces apoptosis and downregulation of major histocompatibility complex class I proteins in tumor T and B cell lines. *Int. J. Oncol.,* 2002 20, 563-570.

[46] Atlante, A et al. Rapid uncoupling of oxidative phosphorylation accompanies glutamate toxicity in rat cerebellar granule cells. *Neuroreport,* 1996 7, 2519-2523.

[47] Atlante, A; Bobba, A; Calissano, P; Passarella, S; Marra, E. The apoptosis/necrosis transition in cerebellar granule cells depends on the mutual relationship of the antioxidant and the proteolytic systems which regulate ROS production and cytochrome c release en route to death. *J. Neurochem.*, 2003 84, 960-971.

[48] Leppa, S; Kajanne, R; Arminen, L; Sistonen, L. Differential induction of Hsp70-encoding genes in human hematopoietic cells. *J. Biol. Chem.*, 2001 276, 31713-31719.

[49] Lodhi, IJ et al. Inhibition of mono-ADP-ribosyltransferase activity during the execution phase of apoptosis prevents apoptotic body formation. *Arch. Biochem. Biophys.*, 2001 387, 66-77.

[50] Huebscher, KJ et al. Protein isoaspartyl methyltransferase protects from Bax-induced apoptosis. *Gene,* 1999 240, 333-341.

[51] Hirashima, Y et al. Transfection of the plasma-type platelet-activating factor acetylhydrolase gene attenuates glutamate-induced apoptosis in cultured rat cortical neurons. *Brain Res.*, 2000 885, 128-132.

[52] Moroni, F. Tryptophan metabolism and brain function: focus on kynurenine and other indole metabolites. *Eur. J. Pharmacol.*, 1999 375, 87-100.

[53] Deng, G; Pike, CJ; Cotman, CW. Alzheimer-associated presenilin-2 confers increased sensitivity to apoptosis in PC12 cells. *FEBS Lett.*, 1996 397, 50-54.

[54] Wolozin, B et al. Participation of presenilin 2 in apoptosis: enhanced basal activity conferred by an Alzheimer mutation. *Science,* 1996 274, 1710-1713.

[55] Galli, C et al. Increased amyloidogenic secretion in cerebellar granule cells undergoing apoptosis. *Proc. Natl. Acad. Sci. U. S. A,* 1998 95, 1247-1252.

[56] Toh, SY; Wang, X; Li, P. Identification of the nuclear factor HMG2 as an activator for DFF nuclease activity. *Biochem. Biophys. Res. Commun.*, 1998 250, 598-601.

[57] Copani, A et al. Growth conditions differentially affect the constitutive expression of primary response genes in cultured cerebellar granule cells. *Neurochem. Res.*, 1995 20, 611-616.

[58] Catania, MV et al. An enhanced expression of the immediate early gene, Egr-1, is associated with neuronal apoptosis in culture. *Neuroscience,* 1999 91, 1529-1538.

[59] Levkovitz, Y and Baraban, JM. A dominant negative inhibitor of the Egr family of transcription regulatory factors suppresses cerebellar granule cell apoptosis by blocking c-Jun activation. *J. Neurosci.*, 2001 21, 5893-5901.

[60] Norton, JD and Atherton, GT. Coupling of cell growth control and apoptosis functions of Id proteins. *Mol. Cell Biol.*, 1998 18, 2371-2381.

[61] Mashima, T; Udagawa, S; Tsuruo, T. Involvement of transcriptional repressor ATF3 in acceleration of caspase protease activation during DNA damaging agent-induced apoptosis. *J. Cell Physiol,* 2001 188, 352-358.

[62] Eymin, B; Dubrez, L; Allouche, M; Solary, E. Increased gadd153 messenger RNA level is associated with apoptosis in human leukemic cells treated with etoposide. *Cancer Res.*, 1997 57, 686-695.

[63] Ten Dijke, P; Goumans, MJ; Itoh, F; Itoh, S. Regulation of cell proliferation by Smad proteins. *J. Cell Physiol,* 2002 191, 1-16.

[64] Srivastava, SP; Kumar, KU; Kaufman, RJ. Phosphorylation of eukaryotic translation initiation factor 2 mediates apoptosis in response to activation of the double-stranded RNA- dependent protein kinase. *J. Biol. Chem.,* 1998 273, 2416-2423.

In: Neuronal Cell Apoptosis Research
Editor: Relif J. Shandling, pp. 139-156
ISBN 1-60021-456-8


*Chapter 6*

# HALLMARKS OF APOPTOTIC-LIKE CELL DEATH IN RESPONSE TO HYPOXIC INJURY IN VARIOUS DEVELOPMENTAL MODELS ARE CLOSELY RELATED TO BRAIN IMMATURITY

***Jean-Luc Daval[1,*] and Christiane Charriaut-Marlangue[2]***

[1]INSERM U.724, Université Henri Poincaré, Faculté de Médecine, 9 avenue de la Forêt de Haye, F-54500 Vandoeuvre-lès-Nancy, France;

[2]Groupe Hypoxie et Ischémie du Cerveau en Développement, Université Pierre et Marie Curie-Paris6, UMR-CNRS 7102, 9 quai St-Bernard, Paris, F-75005 France.

## ABSTRACT

With regard to the specificity of the developing brain, a better understanding of cellular mechanisms involved in perinatal hypoxic-ischemic injury would help to prevent neurological impairments. We therefore examined temporal features of brain injury in three different developmental models of oxygen deprivation capable of inducing apoptotic cell death. Nuclear staining by DAPI (4,6-diamidino-2-phenylindole) was used to identify healthy, apoptotic and necrotic nuclei as well as for cell counting in cultures and brain sections. DNA fragmentation was monitored by in situ terminal dUTP nick end labeling (TUNEL) and electrophoresis on agarose gels. Also, the expression profile of apoptosis-related proteins Bax and Bcl-2 was studied by immunohistochemistry. In all cases, oxygen deprivation induced significant delayed cell death with morphological features of apoptosis and a progressive increase in the Bax/Bcl-2 protein ratio, except in the penumbra of the ischemic infarct where Bcl-2 remained predominant. As in the control newborn brain that still exhibited physiological death, hypoxia-associated DNA

* Correspondence concerning this article should be addressed to Dr. Jean-Luc Daval, INSERM U.724, Université Henri Poincaré, Faculté de Médecine, 9 avenue de la Forêt de Haye, F-54500 Vandoeuvre-lès-Nancy, France. E-mail: Jean-Luc.Daval@nancy.inserm.fr.

breakdown led to small fragments of ~200 bp in the cortex of hypoxic rat pups. Ladder pattern and TUNEL-positive cells exhibiting apoptotic bodies were only present in the penumbra of 7 day-old ischemic rats. These data indicate that hallmarks of hypoxia-induced apoptosis may vary according to brain maturity, possibly through specific nuclease activities. While retaining a part of the developmental death program, the newborn brain seems to be prone to an apoptotic-like response that resembles physiological programmed death.

## INTRODUCTION

Apoptosis is considered as the ubiquitous form of naturally occurring cell death that plays a fundamental role in brain development, as about half of the cells in the immature brain are eliminated by apoptosis (Oppenheim, 1991). This physiological process needs the activation of an intrinsic 'death program' requiring time and energy (Richter et al., 1996) as well as gene transcription and translation (Pittman et al., 1993). Indeed, the apoptotic cascade involves the participation of 'killer' proteins (e.g., Bax) generally present constitutively in the cell but normally repressed by their survival counterparts (e.g., Bcl-2). Morphologically, apoptosis is typified by cell shrinkage, chromatin condensation and subsequent nucleus fragmentation, with a specific pattern of DNA breakdown leading usually to multiple segments of approximately 200 bp in length. By contrast, the second form of cell death, namely necrosis, which has been implicated in cell destruction consecutive to severe trauma, induces cellular swelling, membrane disruption with leakage of cell contents to the extracellular space, and leads to random DNA degradation (Wyllie, 1981). In response to a variety of insults, all cells seem to be able to undergo apoptosis but there is compelling evidence to suggest that developing brain cells are more prone to programmed death by retaining a part of the developmental cell death program (Blaschke et al., 1996; Sidhu et al., 1997).

Perinatal hypoxic-ischemic injury remains a major cause of mortality and cerebral morbidity, susceptible to generate permanent neurological sequelae (Volpe, 1987; Berger and Garnier, 1999). Whereas necrotic cell damage was first considered prevalent in response to hypoxia-ischemia, the participation of apoptosis has been largely documented, including in the developing brain (Beilharz et al., 1995; Charriaut-Marlangue et al., 1996; Bossenmeyer et al., 1998, Bossenmeyer-Pourié et al., 2002; Daval et al., 2004). In addition, some studies suggest an apoptotic-necrotic continuum in the consequences of acute cerebral ischemia (Nakajima et al., 2000; Benchoua et al., 2001).

Since the developing brain displays specific properties and sensitivity to oxygen supply (Grafe, 1994), hallmarks of cell injury may thus vary according to brain maturity, and the present study was designed to gain a better understanding of the cellular mechanisms triggered by a severe hypoxic insult at various ages of the perinatal period. For this purpose, we compared hypoxia-associated features of cell death in three different developmental models of oxygen deprivation capable of inducing brain apoptosis, i.e. in cultured neurons from the embryonic rat forebrain exposed to 95% $N_2$/5% $CO_2$ (Bossenmeyer et al., 1998), in the rat neonate exposed to 100% $N_2$, a model corresponding to birth asphyxia (Grojean et al., 2003), and in the 7-day-old rat pup subjected to transient unilateral focal ischemia followed by reperfusion (Renolleau et al., 1998). In contrast, exposure to 100% $N_2$ in 7 day-old rat induced mainly necrosis and did not enter this study (JLD, unpublished results).

# Experimental Procedures

## Animals

Experiments were conducted in respect to the French (Statement N° 04223) and European Community guidelines for the care and use of experimental animals. Sprague-Dawley female rats (R. Janvier, Le Genest-St-Isle, France) in the proestrus period, as shown by daily vaginal smears, were housed together with males for 24 h, and then maintained during gestation in separate cages under standard laboratory conditions on a 12:12 h light/dark cycle (lights on at 6:00 a.m.) with food and water available ad libitum.

## Hypoxia in Neuronal Cell Cultures

Primary cultured neurons were obtained from the embryonic rat brain as previously described (Bossenmeyer et al., 1998; Chihab et al., 1998; Bossenmeyer-Pourié et al., 1999a; Grojean et al., 2000). Forebrains of 14-day-old embryos were carefully collected, dissected free of meninges and dispersed in culture medium consisting of a mixture of Dulbecco's modified Eagle's medium (DMEM) and Ham's F12 medium (50:50, ICN Pharmaceuticals, Costa Mesa, CA) supplemented with 5% inactivated fetal calf serum (Valbiotech, Paris, France). After centrifugation at 700 g for 10 min, the pellet was dispersed in the same medium and passed through a 46 μm-pore size nylon mesh. Aliquots of the cell suspension were transferred into 35 mm Petri dishes (Falcon, Becton Dickinson, Le Pont-de-Claix, France) precoated with poly-L-lysine in order to obtain a final density of $10^6$ cells/dish. Cultures were then placed at 37°C in a humidified atmosphere of 95% air/5% $CO_2$. The following day, the culture medium was replaced with a fresh hormonally defined serum-free medium corresponding to the DMEM/Ham's F12 mixture enriched with human transferrin (1 mM), bovine insulin (1 mM), putrescine (0.1 mM), progesterone (10 nM), estradiol (1 pM), Na selenite (30 nM), and also containing fibroblast growth factor (2 ng/ml) and epidermal growth factor (10 ng/ml) (Sigma Chemicals, St Louis, MO). Two days later, the culture medium was renewed with serum-free medium in the absence of growth factors.

After 6 days in vitro, neuronal cells were exposed to hypoxia for 6 h by transferring culture dishes to a humidified incubation chamber thermoregulated at 37°C and filled with 95% $N_2$/5% $CO_2$. Cultures were then returned to normoxic atmosphere, at 37°C, whereas control cells were constantly maintained under standard conditions. When used, 1 μM cycloheximide (CHX, Sigma Chemicals) was added to the culture medium prior to hypoxia and removed by changing the medium as soon as reoxygenation began. Cells were studied as a function of time until 96 h post-reoxygenation to assess the effects of hypoxia.

## In Vivo Birth Hypoxia

Between 8 to 24 h after delivery, the litter size was reduced to 10 pups for homogeneity, and 5 neonates were placed for 20 min in a thermostated plexiglas chamber flushed with 100% N2, whereas the remaining 5 pups were taken as matched controls and exposed to 21% O2/79% N2 for the same time. The rate of gas delivery inside the box was calculated to prevent any overpressure, and was fixed to 3 liters/min. Gas in excess was evacuated through a central lengthwise split on the top of the box. The temperature inside the chamber was adjusted to 36°C to maintain body temperature in the physiological range. Following exposure to gas, all rats were allowed to recover for 20 min in normoxic conditions, and they were then returned to their dams. In these conditions, the final rate of mortality in the hypoxia group was 4%, and surviving animals did not display significant suckling problems. Hypoxic and control rats were finally sacrificed by decapitation at various time intervals between 1 day and 13 days post-exposure. Their brains were rapidly collected and immediately frozen in methylbutane previously chilled to -30°C, and stored at -80°C in plastic bags until used. Thereafter, the brains were coated with embedding medium (4% carboxymethylcellulose in water), and cut at -20°C in a cryostat (Reichert Jung, Frigocut 2800, Les Ulis, France) to generate 5-µm coronal sections at the level of the rostral hippocampus, according to the developing rat brain atlas of Sherwood and Timiras (1970), and tissue sections were mounted onto glass slides for subsequent analyses.

## In Vivo 7-Day Ischemia

Ischemia was performed in 7 day-old rats (17-21 g) of both sexes, as previously described (Renolleau et al., 1998). Rat pups were anaesthetized with intraperitoneal injection of chloral hydrate (300 mg/kg). After 15 min, rats were positioned on their back and a median incision was made in the neck to expose left common carotid artery. Rats were then placed on the right side and an oblique skin incision was made between the ear and the eye. After excision of the temporal muscle, the cranial bone was removed from the frontal suture to a level below the zygomatic arch. Then, the left middle cerebral artery, exposed just after its apparition over the rhinal fissure, was coagulated at the inferior cerebral vein level. After this procedure, a clip was placed to occlude the left common carotid artery and was removed after 1 h. Carotid blood flow restoration was verified with the aid of a binocular loupe. Both neck and cranial skin incisions were then closed. During surgery, body temperature was maintained at 37-38°C. After awake, rat pups were transferred to their mother for time survival at 6, 24 and 48 h and 3 to 5 days after reperfusion.

## Evaluation of Cell Damage

Cell viability was measured in cultured neurons by the spectrophotometric method using 3-[4,5-dimethylthiazol-2-yl]-2,5-diphenyltetrazolium bromide (MTT), according to Hansen et al. (1989). Neurons were incubated for 3 h at 37°C with MTT (500 µg/mL, Sigma

Chemicals), washed twice with ice-cold phosphate-buffered saline (PBS), and lysed in dimethyl sulfoxide (DMSO) which solubilizes the residual formazan salt for subsequent quantification. Optical density was measured at 519 nm and data were compared to those obtained from sister control cells to which 100% viability was assigned.

In vivo cell counts were performed to determine the percentage of damaged neurons in cerebral cortex. For this purpose, brain sections were fixed for 10 min in a mixture of ethanol:acetic acid (3:1), washed for 30 seconds in distilled water, air-dried, and then stained for 10 min with the fluorescent dye 4,6-diamidino-2-phenylindole (DAPI, Sigma Chemicals) in phosphate-buffered saline (0.5 μg/ml) (Wolvetang et al., 1994). Sections were washed twice with distilled water, air-dried, and treated with anti-fading medium (10 mg/ml p-phenyldiamine in 90% glycerol, pH 9.0). The number of cell nuclei which were labeled by DAPI was scored at an excitation wavelength of 365 nm under fluorescence microscopy (Zeiss Axioscop, Strasbourg, France). Cell density was measured in the infragranular part of the parietal cortex (layers III-V) at a 40 x magnification in at least 3 separate experiments by counting cells in 3 distinct section areas delineated by an ocular grid of 1/400 mm2. For each selected field, only cells with their nuclei present in the focal plane were counted. Numbers of cells were calculated per mm2 and finally reported as percentages of change from matched controls.

## Monitoring of Apoptosis and Necrosis

Morphological hallmarks of apoptosis and necrosis were analyzed both in cultured neurons and tissue sections after nuclear labeling by DAPI, as previously documented (Bossenmeyer et al., 1998; Chihab et al., 1998; Bossenmeyer-Pourié et al., 1999b; Grojean et al., 2000; Park et al., 1997). Indeed, it has been demonstrated that healthy cells exhibit intact round-shaped nuclei with diffuse fluorescence, indicative of homogeneous chromatin. Necrotic cells are characterized by highly refringent smaller nuclei with uniformly dispersed chromatin, while condensation and fragmentation of chromatin lead to typically shrunken nuclei in apoptotic cells, along with apoptotic bodies. Characteristic nuclei were scored under fluorescence microscopy (Zeiss Axioscop, Strasbourg, France) at an excitation wavelength of 365 nm in at least 3 separate experiments by counting concerned cells in at least 3 distinct areas of 100 cells.

## Electrophoretic Detection of DNA Fragmentation

At 48 h and 96 h post-reoxygenation, hypoxic and control cultured neurons were washed twice with PBS, scraped off in 2 ml PBS, and the contents of 5 dishes were pooled to be centrifuged for 5 min at 700 g. Following in vivo birth hypoxia or ischemia, rat pups were sacrificed at various time intervals and the brains were rapidly removed, dissected on a cold plate and stored at -80° C. Cell pellets or tissue samples from separate brains were processed for DNA isolation according to Laird et al. (1991).They were gently homogenized and lysed in 0.5 ml of lysis buffer (100 mM tris-HCl at pH 8.5, 5 mM EDTA, 0.2% sodium dodecyl sulfate (SDS), and 200 mM NaCl) containing 100 µg/ml proteinase K. After 16 h incubation at 55°C, DNA was precipitated by adding one volume isopropanol with continuous agitation for 15 min at room temperature. Following centrifugation at 13000 g for 5 min the pellet was air dried and dissolved in TE buffer (several hours), treated with RNAse A (20 µg/ml) for 2 h at 37°C and DNA content determined spectrophotometrically. DNA (10 µg/lane) was electrophoresed on a 1% agarose gel in 100 mM tris borate (60 V for 4 h) in the presence of 0.3 mg/ml ethidium bromide and visualized with UV illumination.

## Detection of DNA Breaks by Nick End-Labeling

Sections were processed for DNA strand breaks (TUNEL assay) using the in situ Cell Death Detection Kit, Fluorescein (Roche, Meylan, France) according to the manufacturer's instructions. TUNEL assay reveals apoptosis and necrosis as previously reported (Charriaut-Marlangue and Ben-Ari, 1995).

## Bax and Bcl-2 Immunohistochemistry

The expression of the two prototypic apoptosis-related proteins Bax and Bcl-2 was analyzed in cultured neurons and rat brain coronal sections. In vitro neuronal cells were rinsed twice with PBS, then fixed for 10 min in methanol at -10°C, and rinsed again with PBS. Non-specific binding sites for IgG were blocked by incubating the cells for 20 min with 10% horse serum (Gibco-BRL, Inchinnan, U.K.) in PBS. Thereafter, cultures were incubated for 60 min at room temperature in buffer containing the primary antibody., i.e. a rabbit polyclonal antibody against Bax (N-20, Santa Cruz, Tebu, France) diluted at 1/20 or a goat polyclonal antibody against Bcl-2 (N-19, Santa Cruz) diluted at 1/40. Following two washing steps to remove unfixed antibody, the cells were incubated for 120 min in the presence of a second-step antibody corresponding to either an anti-rabbit IgG conjugated to indocarbocyanine (Cy3, dilution 1/80) or an anti-goat IgG conjugated to rhodamine (TRITC, dilution1/100), both from Jackson ImmunoResearch Laboratories (West Grove, PA). Cultured cells were finally washed 3 times with PBS, coverslipped using mounting medium (Aquapolymount), and kept in the dark until fluorescence analysis by means of a Zeiss Axioscop microscope.

For quantitative analysis, cell fluorescence activity was computerized from microphotographs, and mean intensity was calculated by using Adobe Photoshop® software and expressed as arbitrary units of mean emission per 1 000 pixels (Bossenmeyer-Pourié et al., 2002; Bossenmeyer-Pourié and Daval, 1998). The results were finally reported as Bax/Bcl-2 protein ratios as a function of time after reoxygenation as well as in normoxic control cultures processed in parallel.

Both proteins were also analyzed in rat brain coronal sections of 7 µm-thickness previously fixed by incubating the slides for 10 min at 4°C in acetic acid:ethanol (1:3) in the presence of 30% hydrogen peroxide. In these conditions, brain sections were incubated at 4°C for 48 h with the primary antibody against Bax (dilution 1/20) or Bcl-2 (dilution 1/30), and secondary antibodies were used at 1/50 and 1/100, respectively, for subsequent measurements as described above. In the case of coronal sections from 7 day-old ischemic pups, Bax and Bcl-2 immunoreactivity was visualized by the avidin-biotin peroxidase (Elite ABC kit, Vecstastain, Vector, AbCys, France). The peroxidase activity was evidenced with the use of 3,3'-diaminobenzidin (DAB) and 0.02 % hydrogen peroxide. The counting of Bax- and Bcl-2 positive cells within the core and penumbra was performed in 3-5 sections (at the level of the anterior commissure) using a x40 objective, and results were reported as indicated above.

# RESULTS

## Hypoxia in Neuronal Cell Cultures

As illustrated in Table 1, a hypoxic episode in cultured neurons from the embryonic rat brain induced delayed cell death inasmuch as cell viability was not yet different from controls at 48 h post-reoxygenation. Thereafter, the number of living cells gradually decreased to reach 64% from controls at 96 h. At this experimental time point, a significant number of cell nuclei stained by the fluorescent dye DAPI exhibited characteristic apoptosis-related morphological features, such as condensed chromatin, whereas the presence of apoptotic bodies could be seen. Necrotic cells were also depicted, but the percentage of apoptotic nuclei increased more markedly, as revealed by cell counts. Moreover, a concomitant treatment with cycloheximide, a potent inhibitor of protein synthesis, had beneficial effects. In the presence of CHX, cell viability was not significantly affected and the number of apoptotic nuclei remained within control values (Table 1), suggesting that hypoxia triggers a programmed death process in cultured neurons. When DNA fragmentation was monitored on agarose gel, no significant degradation was observed at 48 h, in good agreement with viability and morphology data, whereas DNA fragmentation was shown at 96 h after the hypoxic insult (Figure 1). However, DNA alteration was only reflected by a smear, without visualization of a 'ladder pattern'. Finally, immunohistochemical studies revealed detectable baseline values of the prototypic apoptosis-related proteins Bcl-2 and Bax. Following exposure to hypoxia, expression of the survival Bcl-2 protein transiently increased above control values at 48 h post-insult without apparent change in Bax expression, leading to a significantly reduced Bax/Bcl-2 ratio (Figure 2). At 96 h, Bcl-2 levels abruptly declined, while the expression of Bax was markedly stimulated, resulting in a robust increase of the final Bax/Bcl-2 ratio (Figure 2).

**Table 1. Effects of a 6-h exposure to hypoxia of cultured rat forebrain neurons on cell viability and nuclear hallmarks of necrosis and apoptosis at 48 h and 96 h post-reoxygenation. Influence of cycloheximide.**

| | Cell viability *(% from controls)* | Necrosis *(% of total neurons)* | Apoptosis *(% of total neurons)* |
|---|---|---|---|
| Controls 48 h | 100.0 ± 4.6 | 4.9 ± 1.2 | 1.6 ± 0.6 |
| + 1 µM CHX | 90.6 ± 5.3 | 9.1 ± 2.3* | 1.2 ± 0.9 |
| Hypoxia 48 h | 96.6 ± 4.3 | 5.2 ± 1.1 | 1.9 ± 0.9 |
| + 1 µM CHX | 92.0 ± 6.1 | 11.0 ± 2.6*°° | 1.3 ± 0.8 |
| Controls 96 h | 100.0 ± 5.8 | 5.6 ± 1.5 | 2.7 ± 0.8 |
| + 1 µM CHX | 86.4 ± 6.0* | 13.4 ± 2.6** | 2.8 ± 1.0 |
| Hypoxia 96 h | 63.6 ± 6.2** | 10.8 ± 2.8** | 21.2 ± 5.7** |
| + 1 µM CHX | 88.5 ± 5.1*°° | 16.1 ± 3.0**° | 3.0 ± 1.2°° |

Cell viability as well as rates of necrosis and apoptosis were analyzed as described in the method section. Data are reported as means ± S.D. obtained from 3 separate experiments. Statistically significant differences from controls: *$p<0.05$ and **$p<0.01$; Statistically significant differences from hypoxia alone: °$p<0.05$ and °°$p<0.01$ (Dunnett's test for multiple comparisons).

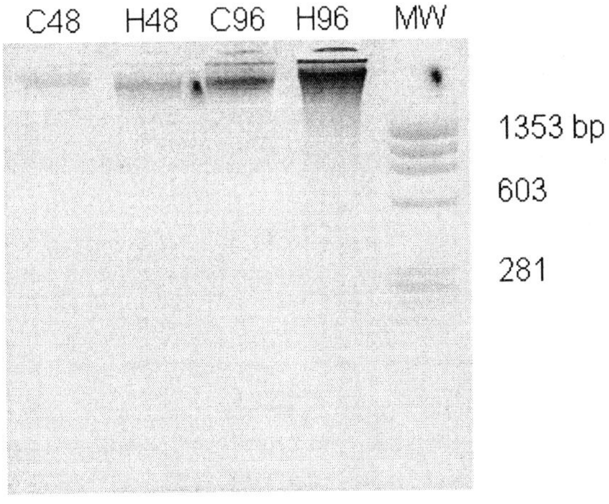


Figure 1. Temporal analysis on agarose gel of DNA isolated from control and hypoxic cultured neurons from the embryonic rat forebrain. DNA was isolated from cultured cells at 48 h (H48) and 96 h post-reoxygenation (H96) and from matched controls (C48 and C96). One representative example is presented for illustration, and similar profiles were obtained from 5 separate experiments. (Size marker: Superladder mid 1-500 bp, Eurogentec, Paris, France).

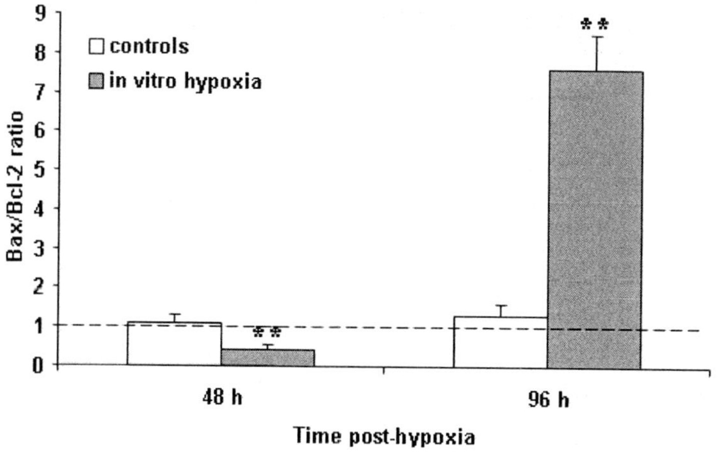


Figure 2. Evolution of the Bax/Bcl-2 protein ratio in cultured neurons exposed to hypoxia for 6 hours and their normoxic controls. Immunohistochemical studies were performed at 48 h and 96 h post-hypoxia and in matched controls. Fluorescence activity was computerized by means of Adobe Photoshop software, and the protein ratio was calculated at each time point. The data represent mean values (± S.D.) obtained from 3 separate experiments. Statistically significant difference from controls: ** $p<0.01$ (Dunnett's test for multiple comparisons).

## In Vivo Birth Hypoxia

When rat pups were subjected to hypoxia within 24 h after birth, no histopathological alteration of the cerebral cortex could be noticed by the four ensuing days. Then, a progressive decline of the total number of cells labeled by DAPI was observed in this brain structure, and maximal cell loss was seen around 6 days post-reoxygenation (Table 2). Whereas examination of nuclear morphology depicted a significant number of cells exhibiting the characteristic hallmarks of both necrosis and apoptosis during early postnatal life in control animals, birth hypoxia was associated with a gradual elevation of the percentage of affected cells, with a peak observed at 6 days post-reoxygenation. The number of altered nuclei was then shown to decrease during the second week after hypoxia, and only the proportion of necrotic cells remained significantly augmented by comparison with controls at 13 days after the insult (Table 2). Surprisingly, no TUNEL-positive nuclei were detected on coronal sections tested at various time intervals after hypoxia (not shown). As illustrated in Figure 3, where only representative time-points are presented, analysis of DNA on agarose gel revealed similar fragmentation in control and hypoxic rat cortex during the first 3 days following exposure to gas. In these conditions, only one spot could be repeatedly visualized, and the size marker indicated that such fragments corresponded to ~200 bp. Between 5 and 7 days post-exposure, DNA fragmentation was exclusively seen in the cortex of hypoxic rats, with the absence of alteration in controls. Again, hypoxia led to DNA fragments of ~200 bp. At 13 days post-insult, no DNA fragmentation could be observed.

**Table 2. Temporal evolution of the number of cells and their nuclear morphological features of necrosis and apoptosis in the cerebral cortex of rats exposed to birth hypoxia for 20 min.**

| Time post-reoxygenation | Number of total cells *(% from controls)* | Necrosis (%) | | Apoptosis (%) | |
|---|---|---|---|---|---|
| | | Controls | Hypoxia | Controls | Hypoxia |
| 1 day | 98.3 ± 5.7 | 5.3 ± 0.5 | 4.8 ± 1.7 | 6.7 ± 1.3 | 4.8 ± 1.7 |
| 2 days | 98.3 ± 1.5 | 3.2 ± 1.0 | 2.9 ± 1.0 | 4.3 ± 0.9 | 4.7 ± 0.8 |
| 3 days | 95.6 ± 1.7 | 1.4 ± 1.2 | 1.8 ± 1.2 | 2.2 ± 0.8 | 2.2 ± 0.6 |
| 4 days | 86.5 ± 3.6** | 1.2 ± 0.1 | 3.6 ± 0.8** | 2.4 ± 0.7 | 6.0 ± 0.5** |
| 5 days | 80.6 ± 6.2** | 0.8 ± 1.1 | 4.7 ± 0.6** | 2.0 ± 0.3 | 6.0 ± 0.5** |
| 6 days | 77.1 ± 9.2** | 2.7 ± 0.8 | 8.7 ± 0.7** | 3.1 ± 0.1 | 9.0 ± 0.6** |
| 7 days | 78.7 ± 7.1** | 1.3 ±1.1 | 4.3 ± 0.1** | 2.2 ± 0.2 | 6.9 ± 0.6** |
| 13 days | 78.5 ± 7.3** | 3.7 ±0.6 | 5.6 ± 0.3* | 2.9 ± 0.6 | 3.1 ± 0.5 |

Total number of cells was measured in the cerebral cortex as a function of time after hypoxia by cell counting following nuclear staining by DAPI and is expressed as a percentage from matched controls. Proportions of necrotic and apoptotic nuclei were calculated after DAPI labeling according to morphological hallmarks and are expressed as a percentage of total cells in both control and hypoxic rats. Data were obtained from 3 to 5 separate experiments using animals from different litters, and are given as means ± S.D. Statistically significant differences from controls: *$p<0.05$ and **$p<0.01$; Statistically significant differences from hypoxia alone: °$p<0.05$ and °°$p<0.01$ (Dunnett's test for multiple comparisons).

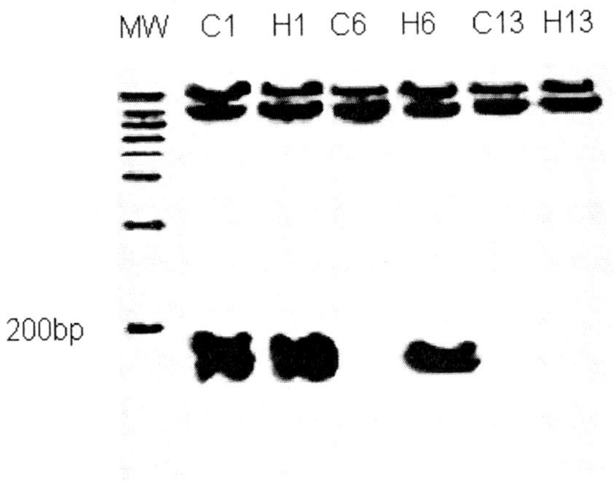


Figure 3. Temporal analysis on agarose gel of DNA isolated from the cerebral cortex of control and hypoxic rats. Cortical DNA was isolated at 1 day (H1), 6 days (H6) and 13 days (H13) after birth hypoxia and from matched controls (C1, C6 and C13). One representative example is presented for illustration, and similar profiles were obtained from 5 separate experiments.

Regarding Bax and Bcl-2, a significant basal expression of both proteins was detected in the cortex of control animals of 1-4 days of age, and their levels tended to decrease thereafter. The resulting Bax/Bcl-2 ratio was globally maintained around 1 during the period examined (Figure 4). Following birth hypoxia, Bcl-2 was transiently overexpressed at 3 days post-insult, and its expression gradually declined as a function of time, to be reduced by 38% by comparison to matched controls at 13 days. Bax expression was similar to controls at 3 days, and then markedly increased in response to birth hypoxia, depicting a significant elevation of the Bax/Bcl-2 protein ratio at the final stage (Figure 4).

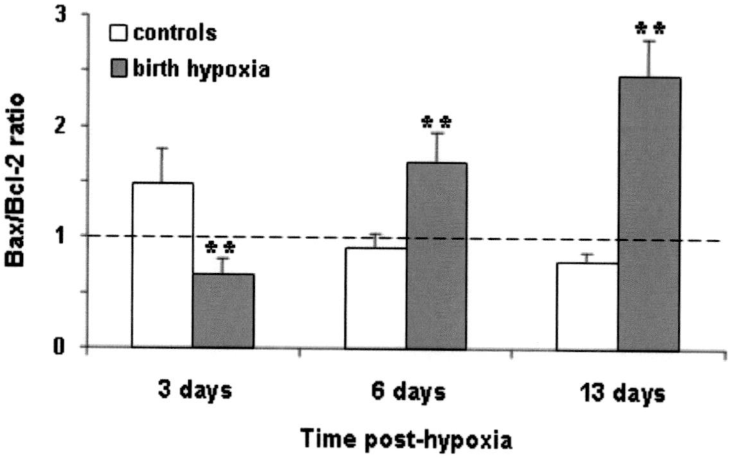


Figure 4. Evolution of the Bax/Bcl-2 protein ratio in the cerebral cortex of rats exposed to birth hypoxia and their matched controls. Immunohistochemical studies were performed at 3 days, 6 days and 13 days post-hypoxia and in control rat pups. Fluorescence activity was analysed as described in the legend of Figure 2. The data represent mean values (± S.D.) obtained from 3 separate experiments. Statistically significant difference from controls: ** $p<0.01$ (Dunnett's test for multiple comparisons).

## In Vivo 7-Day Ischemia

When 7-day old rat pups were subjected to ischemia, a cortical infarct was clearly delineated at 48 h of reperfusion, as previously documented (Renolleau et al., 1998). At this time point of recovery, a decline in the number of cells was recorded. Three zones were depicted, the core of the infarct, penumbra close to the core and distant penumbra in which 13, 34 and 52 % of surviving cells were respectively counted, as compared to the contralateral cortex (Table 3). With DAPI staining and examination of the nuclear morphology, cells exhibiting the characteristic hallmarks of apoptosis were significantly detected in the 2 penumbral zones, whereas nuclei with a necrotic profile were mainly observed in the core of the infarct (Table 3). Similar findings were obtained by the TUNEL assay. Whereas the presence of necrotic nuclei with flocculent chromatin was shown in the core (Figure 5 A, B), all of the TUNEL-positive nuclei in the penumbra demonstrated chromatin condensation and apoptotic body formation (Figure 5 C). Analysis of DNA fragmentation on agarose gel demonstrated DNA laddering in the ipsilateral cortex, which mainly corresponded to penumbral tissue, 5 days after the onset of ischemia, which was not present in the contralateral cortex (Figure 5D).

**Table 3. Temporal evolution of the number of cells and their nuclear morphological features of necrosis and apoptosis in the cerebral cortex of 7 day-old rats exposed to ischemia and killed at 48 h of reperfusion.**

| | **Number of total cells** *(per 0.01 mm$^2$)* | **Necrosis** *(% of total cells)* | **Apoptosis** *(% of total cells)* |
|---|---|---|---|
| Contralateral cortex | 72.0 ± 1.4 | 2.1 ± 0.5 | 0.7 ± 0.1 |

**Table 3 (continued)**

| | **Number of total cells** *(per 0.01 mm$^2$)* | **Necrosis** *(% of total cells)* | **Apoptosis** *(% of total cells)* |
|---|---|---|---|
| Ischemic Core | 12.3 ± 1.1** | 82.3 ± 1.9** | 13.3 ± 0.5** |
| Ischemic Penumbra (a) | 31.5 ± 0.7** | 6.3 ± 0.3** | 58.1 ± 0.2**□ |
| Ischemic Penumbra (b) | 48.5 ± 0.7** | 4.2 ± 0.1 | 30.6 ± 0.5** |

Cell density as well as rates of necrosis and apoptosis were analyzed as described in the method section. Data are reported as means ± S.D. obtained from 3 separate experiments. Statistically significant differences from contralateral cortex (control): **$p<0.01$ (Dunnett's test for multiple comparisons). Penumbra (a) = close to the core; Penumbra (b) = distant from the core.

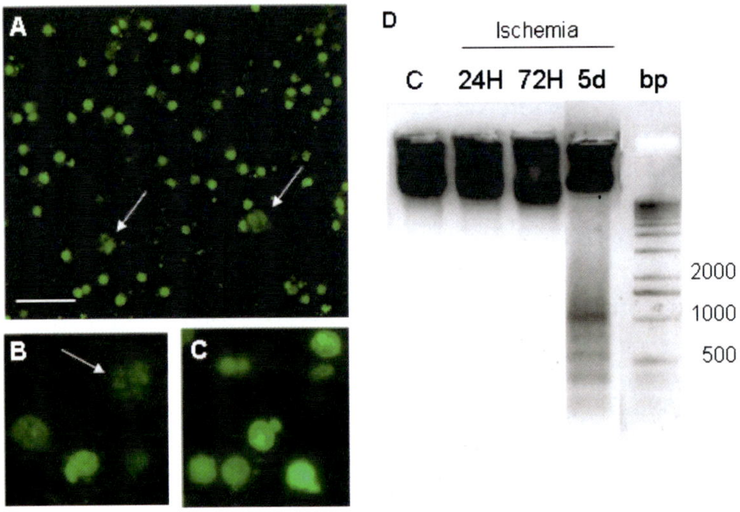


Figure 5. Typical illustration of DNA fragmentation by the TUNEL assay and agarose gel of genomic DNA from ischemic rat pups. A-C: Nuclei exhibited flocculent chromatin in the core (white arrows in A and B) and chromatin condensation and apoptotic bodies in both core (A, B) and penumbra (C). D: Temporal analysis on agarose gel of DNA isolated from ipsilateral (i) and contralateral (c) cortex. Cortical DNA was isolated at 48 h (H48), 4 days (D4) and 5 days (D5) post-ischemia and compared to control (C). Note DNA fragmentation with formation of a ladder at 5 days post-injury. One representative example is presented for illustration, and similar profiles were obtained from 4 separate experiments. Bar represents 100 (A) and 20 μm (B, C).

A basal Bax immunoreactivity was observed in a few cells scattered in the control cortex and in the contralateral side of ischemic brains whereas no Bcl-2-positive cells was detected. Quantification of the number of Bax-positive cells indicated a gradual elevation of their number after reperfusion in the core of the infarct in which a slight number of Bcl-2-positive cells were present. The resulting Bax/Bcl-2 ratio progressively increased between 6 and 24 h and was conspicuous at 48 h of reperfusion (Figure 6) in the core of the infarct. In contrast, Bcl-2-positive cells were still detected in the penumbra (close and distant from the core), leading to a similar Bax/Bcl-2 ratio at 24 h. Cell death in the penumbra began to be effective, as the Bax/Bcl-2 ration increased, at 48 h of recovery (Figure 6).

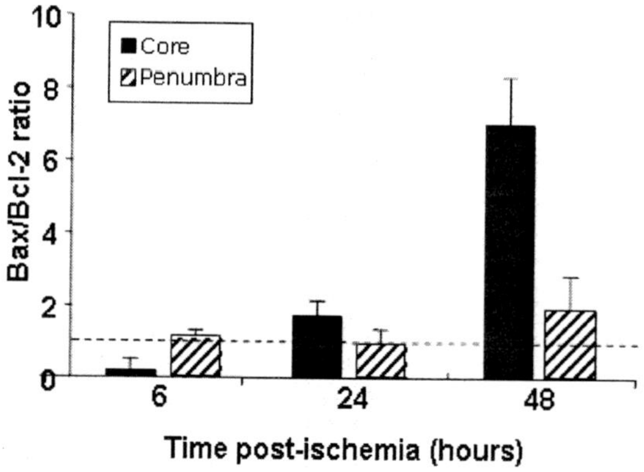


Figure 6. Evolution of the Bax/Bcl-2 positive cell ratio in the cerebral cortex of 7-day-old rats exposed to ischemia. Immunohistochemical studies were performed at 6, 24 and 48 h post-ischemia. No Bcl-2 positive cells and a few (2 to 4) Bax positive cells have been detected in controls. The data represent mean values ± S.D. obtained from 3 separate experiments.

## CONCLUSION

During development, apoptosis is a crucial event for maintaining the brain homeostasis and setting up appropriate neuron-target cell connections (Oppenheim, 1991). In the rat brain, the physiological death program does not take place exclusively during embryogenesis but is also present during early postnatal development. Consistently, characteristic features of programmed cell death were observed during the life span of cultured neurons isolated from the forebrain of 14-day-old embryos as well as in the brains of rat pups by the first days of postnatal life. For example, baseline levels of Bax and Bcl-2 proteins were routinely detected in control samples, and higher Bax levels depicted by the first days of postnatal age are indicative of active apoptosis in the rat brain, although other members of the Bcl-2 family may play a critical role (Isenmann et al., 1998; Otter et al., 1998). In good agreement, the numbers of DAPI-labeled nuclei exhibiting morphological features of condensed chromatin were more elevated in the cortex of rats at 1 and 2 days of age. By contrast to the newborn brain, Bcl-2 was not detectable at baseline in the cerebral cortex of 7-day-old rats. Indeed, it is known that the expression of Bcl-2 in rodents is developmentally regulated, with highest levels present in the antenatal brain (Abe-Dohmae et al., 1993) where it would participate to the selection of surviving neurons during the period of physiological death. Finally, when DNA fragmentation was monitored by electrophoresis, significant amounts of small fragments corresponding to ~200 bp were repeatedly observed, in the absence of typical ladder pattern, in the cortex of control rats until 3 days of age, before to disappear in older animals. According to our criteria, necrotic cells were also detected in cultured neurons and even in the cortex of rat pups in the early postnatal period. It is known, however, that necrosis may appear as an ultimate feature of apoptosis (Nakajima et al., 2000; Benchoua et al., 2001), especially in culture condition where dying cells are not submitted to active phagocytosis (Bonfoco et al., 1995). All the above observations confirm the high propensity of developing brain cells to undergo apoptosis.

The participation of apoptotic-like cell death has been widely described in various pathological states, including in post-mitotic neurons (see Sastry and Rao, 2000 for review). Exposure to hypoxia elicited delayed cell death in the three models examined. Programmed death was confirmed by the beneficial effects of cycloheximide in cultured neurons, as previously documented (Bossenmeyer et al., 1998; Papas et al., 1992), and was shown to require activation of caspases, especially caspase-3 in the various models (Bossenmeyer-Pourié et al., 1999; 2000; Grojean et al., 2003; Benjelloun et al., 2003). Also, apoptotic death is known to be associated with an elevated Bax/Bcl-2 ratio within the cells (Otter et al., 1998). In this respect, Bcl-2 was transiently predominant at 48 h post-insult in cultured neurons. Then, Bcl-2 level decreased, in parallel with a substantial stimulation of Bax expression. By the same time, the presence of typically shrunken nuclei and a reduced number of viable cells were recorded. Such a temporal profile of protein expression was also depicted in the cortex of the asphyxiated newborn rat as well as in the core of the infarct following ischemia in the 7-day-old rat, and these findings are consistent with the view of an apoptotic-necrotic continuum in response to severe neuronal injury (Martin, 2001). In the penumbral area, a long-lasting stimulation of Bcl-2 was recorded, probably due to more effective homeostatic mechanisms and higher availability of energy compared to the core of the infarct where necrosis developed more rapidly.

In the present study, DAPI staining, TUNEL assay and gel electrophoresis of extracted DNA were used to investigate alterations of nuclear DNA in response to hypoxia. Although chromatin condensation and formation of apoptotic bodies were visualized by DAPI in all cases, DNA fragmentation was demonstrated by the TUNEL assay following ischemia in 7 day-old rat pups but not after birth hypoxia. Similarly, DNA laddering was found in the former model but not in the latter, except fragments of ~200 bp. These results suggest that DNA fragmentation occurred in a different manner, probably dependent on the age of injured animals. DNA digestion at internucleosomal linker regions was recognized in 1980 as a characteristic feature of apoptosis, resulting in small double-stranded fragments of DNA that migrate in a ladder pattern (Wyllie, 1980). The fragmentation of ladder type has been attributed to the activities of $Ca^{2+}/Mg^{2+}$-activated endonucleases (Kyprianou et al., 1988), DNAse I (Arends et al., 1990) or DNAse II (Barry ane Eastmann, 1993). However, less extensive digestion of DNA to sizes of 50 and 300 kbp, corresponding to the loop and rosette structures of higher-order chromatin, occurs in several examples of apoptosis without or prior to processing to oligonucleosome-sized fragments and can be recognized by fractionation of genomic DNA on pulse field gel electrophoresis (Walker et al., 1991; Brown et al., 1993; Oberhammer at al., 1993). Such high molecular weight fragments have been shown following ischemia in the P7 rat pup as early as 4 h of recovery (Charriaut-Marlangue et al., 1999). They appeared prior to the detection of TUNEL-positive nuclei and DNA laddering, and have been previously reported in the brain of P7 rat subjected to the Rice model of hypoxia-ischemia (Hou et al., 1997). Taken together, our results suggest that birth hypoxia may induce high molecular weight DNA fragments which cannot be detected on agarose gels, and 200 bp fragments were probably the result of an endogenous exonuclease activity. When neuronal cultures were submitted to hypoxia, a DNA smear was observed on agarose gels. Consistently, serum deprivation of PC12 cells was reported to induce apoptosis along with a DNA degradation as a smear which was prevented by aurintricarboxylic acid, an endonuclease inhibitor (Dessi et al., 1993). Furthermore, there is at least one in vitro example of apoptotic death in which DNA digestion is not detectable by sensitive methods, consistent with experimental evidence that nuclear events may not be required for apoptosis (Ucker et al., 1992). Finally, it is noteworthy that no DNA degradation could be seen at 13 days after birth asphyxia, along with the lack of apoptotic nuclei as depicted by DAPI staining. Such observations suggest the possibility of repair processes, in line with the remarkable plasticity of the juvenile brain.

In conclusion, oxygen deprivation triggered delayed cell death that mainly corresponds to apoptosis, according to morphological and biochemical criteria, in the three experimental models investigated in the present study. The consequences of hypoxia in cultured neurons, which were shown to be dependent upon protein neosynthesis, are comparable to those observed in the newborn rat brain, these two models being theorically matched with regard to their developmental stages (Dichter, 1978). Furthermore, we have shown that when it was monitored by 4-6 days post-reoxygenation, DNA fragmentation elicited by birth asphyxia exhibited the same characteristics as DNA fragmentation consecutive to naturally occurring cell death in the newborn brain (i.e., TUNEL-negative in situ and leading solely to small fragments of approximately 200 bp on agarose gels). Conversely, hypoxia-ischemia in the 7-day-old rat induced apototic cell death in the penumbra with characteristics closer to those reported in the adult brain, but with reduced level of necrosis compared to the adult (Li et al., 1998). Therefore, our results suggest that a shift may occur in the rat brain by the end of the first week of postnatal age with respect to the hallmarks of apoptotic-like cell injury in response to hypoxia, with brain manifestations in the neonate mainly corresponding to the developmental death pattern. These findings might account, at least partly, for the relative resistance of the immature brain to hypoxic insults.

## REFERENCES

Abe-Dohmae, S., Harada, N., Yamada, K., Tanaka, R. Bcl-2 gene is highly expressed during neurogenesis in the central nervous system. *Biochem Biophys Res Commun*, 1993, 191, 915-921.

Arends, M.J., Morris, R.G., Wyllie, A.H. Apoptosis. The role of the endonuclease. *Am J Pathol*, 1990, 136, 593-608.

Barry, M.A. and Eastmann, A. Identification of deoxyribonuclease II as an endonuclease involved in apoptosis. *Arch Biochem Biophys*, 1993, 300, 440-450.

Beilharz, E.J., Williams, C.E., Dragunow, M., Sirimanne, E.S., Gluckman, P.D. Mechanisms of delayed cell death following hypoxic-ischemic injury in the immature rat: evidence for apoptosis during selective neuronal loss. *Mol Brain Res*, 1995, 29, 1-14.

Berger, R., and Garnier, Y. Pathophysiology of perinatal brain damage. *Brain Res Rev*, 1999, 30, 107-134.

Benchoua, A., Guégan, C., Couriaud, C., Hosseini, H., Sampaïo, N., Morin, D., Onténiente, B. Specific caspase pathways are activated in the two stages of cerebral infarction. *J Neurosci*, 2001, 21, 7127-7134.

Benjelloun, N., Joly, L.M., Palmier, B., Plotkine, M., Charriaut-Marlangue, C. Apoptotic mitochondrial pathway in neuron and astrocyte after neonatal hypoxia-ischemia in the rat brain. *Neuropathol & Applied Neurobiol*, 2003, 29, 350-360.

Blaschke, A.J., Staley, K., Chun, J. Widespread programmed cell death in proliferative and postmitotic regions of the fetal cerebral cortex. *Development*, 1996, 122, 1165-1174.

Bonfoco, E., Kraine, D., Ankarcrona, M., Nicotera, P., Lipton, S.A. Apoptosis and necrosis: two distinct events induced, respectively, by mild and intense insults with N-methyl-D-aspartate or nitric oxide/superoxide in cortical cell cultures. *Proc Natl Acad Sci (USA)*, 1995, 92, 7162-7166.

Bossenmeyer, C., Chihab, R., Muller, S., Schroeder, H., Daval, J.L. Hypoxia/reoxygenation induces apoptosis through biphasic induction of protein synthesis in central neurons. *Brain Res*, 1998, 787, 107-116.

Bossenmeyer-Pourié C., Daval, J.L. Prevention from hypoxia-induced apoptosis by preconditioning: a mechanistic approach in cultured neurons from fetal rat forebrain. *Mol Brain Res*, 1998, 58, 237-239.

Bossenmeyer-Pourié, C., Chihab, R., Schroeder, H., Daval, J.L. Transient hypoxia may lead to neuronal proliferation in the developing mammalian brain: from apoptosis to cell cycle completion. *Neuroscience*, 1999a, 91, 221-231.

Bossenmeyer-Pourié, C., Koziel, V., Daval, J.L. CPP32/CASPASE-3-like proteases in hypoxia-induced apoptosis in developing brain neurons, *Mol. Brain Res*, 1999b 71, 225-237.

Bossenmeyer-Pourié, C., Koziel, V., Daval, J.L. Involvement of caspase-1 proteases in hypoxic brain injury. Effects of their inhibitors in developing neurons, *Neuroscience*, 2000, 95, 1157-1165.

Bossenmeyer-Pourié, C., Lièvre V, Grojean S, Koziel, V., Pillot T., Daval, J.L. Sequential expression patterns of apoptosis- and cell cycle-related proteins in neuronal response to severe or mild transient hypoxia. *Neuroscience*, 2002, 114, 869-882.

Brown, D., Sun, X., Cohen, G. Dexamethasone-induced apoptosis involves cleavage of DNA to large fragments prior to internucleosomal fragmentation. *J Biol Chem*, 1993, 268, 3037-3039.

Charriaut-Marlangue, C., Ben-Ari, Y. A cautionary note on the use of TUNEL stain to describe apoptosis, *NeuroReport*, 1995, 7, 61-64.

Charriaut-Marlangue, C., Aggoun-Zouaoui, D., Represa, A., Ben-Ari, Y. Apoptotic features in ischemia, epilepsy and gp 120 toxicity. *Trends in Neurosci*, 1996, 19, 109-114.

Charriaut-Marlangue, C., Richard, E., Ben-Ari, Y. DNA damage and DNA damage-inducible protein Gadd45 following ischemia in the P7 neonatal rat. *Dev Brain Res*, 1999, 116, 133-140.

Chihab, R., Ferry, C., Koziel, V., Monin, P., Daval, J.L. Sequential activation of activator protein-1-related transcription factors and JNK protein kinases may contribute to apoptotic death induced by transient hypoxia in developing brain neurons. *Mol Brain Res*, 1998, 63, 105-120.

Daval, J.L., Pourié, G., Grojean, S., Lièvre, V., Strazielle, C., Blaise, S., Vert, P. Neonatal hypoxia triggers transient apoptosis followed by neurogenesis in the rat CA1 hippocampus. *Pediatr Res*, 2004, 55, 561-567.

Dessi, F., Charriaut-Marlangue, C., Khrestchatisky, M., Ben-Ari, Y. Glutamate-induced neuronal death is not a programmed cell death in cerebellar culture. *J Neurochem*, 1993, 60, 1953-1955.

Dichter, M.A. Rat cortical neurons in cell culture: culture methods, cell morphology, electrophysiology and synapse formation. *Brain Res*, 1978, 149, 279-293.

Grojean, S., Koziel, V., Vert, P., Daval, J.L. Bilirubin induces apoptosis via activation of NMDA receptors in developing rat brain neurons. *Exp Neurol*, 2000, 166, 334-341.

Grojean, S., Schroeder, H., Pourié, G., Charriaut-Marlangue, C., Koziel, V., Desor, D., Vert, P., Daval, J.L. Histopathological alterations and functional brain deficits after transient hypoxia in the newborn rat pup: a long term follow-up. *Neurobiol Dis*, 2003, 14, 265-278.

Grafe MR. Developmental changes in the sensitivity of the neonatal rat brain. *Brain Res*, 1994, 653,161-166.

Hansen, M.B., Nielsen, S.E., Berg, K. Re-examination and further development of a precise and rapid dye method for measuring cell growth/cell kill. *J Immunol Meth*, 1989, 119, 203-210.

Hou, S., Tu, Y., Buchan, A., Huang, Z., Preston, E., Rasquinha, I., Robertson, G., MacManus, J. Increase in DNA lesions and the DNA damage indicator Gadd45 following transient ischemia. *Biochem Cell Biol*, 1997, 75,383-392.

Isenmann, S., Stoll, G., Schroeter, M., Krajewski, S., Reed, J.C., Bahr, M. Differential regulation of Bax, Bcl-2, and Bcl-X proteins in focal cortical ischemia in the rat. *Brain Pathol*, 1998, 8, 49-63.

Kyprianou, N., English, H.F., Isaacs, J.T. Activation of a Ca2+-Mg2+-dependent endonuclease as an early event in castration-induced prostatic cell death. *Prostate*, 1988, 13, 103-117.

Laird, P., Zijderveld, A., Linders, K., Rudnicki, M., Jaenisch, R., Berns, A. Simplified mammalian DNA isolation procedure. *Nucleic Acids Res*, 1991, 19, 4293.

Li, Y., Powers, C., Jiang, N., Chopp, M. Intact, injured, necrotic and apoptotic cells after focal cerebral ischemia in the rat. *J Neurol Sci*, 1998, 156, 119-132.

Martin, L.J. Neuronal cell death in nervous system development, disease, and injury. *Int J Mol Med*, 2000, 7, 455-478.

Nakajima, W., Ishida, A., Lange, M.S., Gabrielson, K.L., Wilson, M.A., Martin, L.J., Blue, M.E., Johnston, M.V. Apoptosis has a prolonged role in the degeneration after hypoxic ischemia in the newborn rat. *J Neurosci*, 2000, 20, 7994-8004.

Oberhammer, F., Wilson, J., Dive, C., Morris, I., Hickman, J., Wakeling, A., Walker, P., Sikorska, M. Apoptotic death in epithelial cells: cleavage of DNA to 300 and/or 50 kb fragments prior to or in the absence of internucleosomal fragmentation. *EMBO J*, 1993, 12, 3679-3684

Oppenheim, R.W. Cell death during development of the nervous system. *Ann Rev Neurosci*, 1991, 14, 453-501.

Otter, I., Conus, S., Ravn, U., Rager, M., Olivier, R., Monney, L., Fabbro, D., Borner, C. The binding properties and biological activities of Bcl-2 and Bax in cells exposed to apoptotic stimuli. *J Biol Chem*, 1998, 273, 6110-6120.

Papas, S., Crépel, V., Hasboun, D., Jorquera, I., Chinestra, P., Ben-Ari, Y. Cycloheximide reduces the effects of anoxic insult in vivo and in vitro. *Eur J Neurosci*, 1992, 4, 758-765.

Park, D.S., Morris, E.J., Greene, L.A., Geller, H.M. G1/S cell cycle blockers and inhibitors of cyclin-dependent kinases suppress camptothecin-induced neuronal apoptosis. *J Neurosci*, 1997, 17, 1256-1270.

Pittman, R.N., Wang, S., DiBenedetto, A.J., Mills, J.C. A system for characterizing cellular and molecular events in programmed neuronal cell death. *J Neurosci*, 1993, 13, 3669-3680.

Renolleau, S., Aggoun-Zouaoui, D., Ben-Ari, Y., Charriaut-Marlangue, C. A model of transient unilateral focal ischemia with reperfusion in the P7 neonatal rat: morphological changes indicative of apoptosis. *Stroke*, 1998, 29, 1454-1461.

Richter C, Schweizer M, Cossarizza A, Franceschi C. Control of apoptosis by the cellular ATP level. *FEBS Lett*, 1996, 378:107-110.

Sastry, P.S., Rao, K.S. Apoptosis and the central nervous system. *J Neurochem*, 2000, 74, 1-20.

Sherwood, N.M., Timiras, P.S. A stereotaxic atlas of the developing rat brain. Berkeley: *University of California Press*, 1970

Sidhu RS, Tuor UI, Del Bigio MR. Nuclear condensation and fragmentation following cerebral hypoxia-ischemia occurs more frequently in immature than older rats. *Neurosci Lett*, 1997, 223, 129-132.

Ucker, D., Obermiller, P., Eckhart, W., Apgar, J., Berger, N., Meyers, J. Genome digestion is a dispensable consequence of physiological cell death mediated by cytotoxic T lymphocytes. *Mol Cell Biol*, 1992, 12, 3060-3069.

Volpe, J.J. Neurology of the newborn. In: Volpe J.J. (Ed) *Neurology of the Newborn.* WB Saunders, Philadelphia, 1987, 236-280.

Walker, P., Smith, C., Youdale, T., Leblanc, J., Whitfield, J., Sikorska, M. Topoisomerase II-reactive chemotherapeutic drugs induce apoptosis in thymocytes. *Cancer Res*, 1991, 51, 1078-1085.

Wolvetang, E.J., Johnson, K.L., Krauer, K., Ralph, S.J., Linnane, A.W. Mitochondrial respiratory chain inhibitors induce apoptosis. *FEBS Letters*, 1994, 339, 40-44.

Wyllie, A.H. Glucocorticoid-induced thymocyte apoptosis is associated with endogenous endonuclease activation. *Nature*, 1980, 284, 555-556.

Wyllie, A.H. Cell death: a new classification separating apoptosis from necrosis. In: Bowen ID, Lockshin RA (eds) *Cell Death in Biology and Pathology*. London: Chapman and Hall, 1981, 9-34.

In: Neuronal Cell Apoptosis Research
Editor: Relif J. Shandling, pp. 157-178
ISBN 1-60021-456-8


*Chapter 7*

# ENVIRONMENTAL EXPOSURES CAN AFFECT APOPTOSIS IN A DEVELOPING BRAIN

***Nazan Uysal* and Osman Acikgoz***
Department of Physiology, University of Dokuz Eylul, Izmir, TURKEY

## ABSTRACT

Apoptosis, known as programmed cell death, is a form of cell death that serves to eliminate dying cells in proliferating or differentiating cell populations, thus plays a crucial role in normal growth and development of many tissues including neurons and tissue homeostasis. Brain development begins in intrauterine period and continues until the end of the adolescence. Developing brain is much more sensitive to environmental exposures such as maternal deprivation, stress, ethanol, head trauma, exercise. Apoptosis plays an important role in developing brain and also environmental exposures can cause apoptosis or increase apoptosis.

Many neuroprotective agents such as melatonin, erythropoietin, MK-801 may prevent environmental exposures-induced apoptosis in developing brains. Approximately on postnatal day 4, the infant rat enters stress-hyporesponsive period (SHRP). A single episode of maternal deprivation for 24h at post-SHRP induce neuronal loss in hippocampus regions of the brain in the infant rats, while it is not influence hippocampal neurons in SHRP. Environmental exposures such as stress, not affect hippocampal neurons in this period. Infant brain might be protected in SHRP from stress-induced apoptosis of brain. Also, head trauma activated physiological apoptotic mechanisms in the developing brains. The immature brain is exceedingly vulnerable to apoptotic neurodegeneration triggered by head trauma and ethanol intake. The effects of exercise opposite of these exposures, increase cell density without altering of apoptosis in the hippocampus and dentate gyrus of adolescent rats. Hippocampal neurons normally play a crucial role in the processing of spatial memory and learning. Memory function, in particular, is at risk following environmental exposures-induced apoptosis due to the selective vulnerability of the hippocampus. To keep a constant number of cells in tissues,

* tel:+90-232-4124460; fax:+90-232-2514606; email: nazan.uysal@deu.edu.tr

there has to be a balance between new cell formation (or proliferation) and cell death (apoptosis). Hippocampal neuron loss is associated with memory deficits following apoptosis. Likewise, increase of hippocampal neurons is closely related to increase in learning and memory.

Apoptosis is an important process in the development of the nervous system, and also apoptosis may be alterable extrinsic factors.

## INTRODUCTION

Apoptosis is a form of cell death that serves to eliminate dying cells in proliferating or differentiating cell populations [1], thus plays an essential process for normal neuronal development of mammalian brain [2,3]. Nearly half of the neurons produced during neurogenesis die before completion of central nervous system maturation via apoptosis.

Apoptosis can be triggered by several stimulus from outside or inside of the cell, such as ligation of a membrane receptor complex at cell surface, by DNA damage as a cause of defects in DNA repair mechanisms, treatment with cytotoxic drugs or irradiation, by a lack of survival signals, contradictory cell cycle signals or by developmental death signals. A family of proteins known as caspases are activated in the early stages of apoptosis. These proteins breakdown or cleave key cellular substrates that are required for normal cellular function including structural proteins in the cytoskeleton and nuclear proteins such as DNA repair enzymes. The caspases can also activate other degradative enzymes such as DNAses. The result of these events is become cell apoptosis [4,5].

Some of apoptotic stimulus is activated mitochondrial proapoptotic events, a cytosolic death signal protein complex that is formed upon release of cytochrome c from the mitochondria [6]. The mitochondrion is the most common source of reactive oxygen species (ROS) during apoptosis [7]. Increased ROS production results in oxidative stress. Oxidative stress is recognized as a strong mediator of apoptosis via the formation of lipid hydroperoxides. Lipid hydroperoxides are well-known to be highly toxic and to induce DNA fragmentation [8,9]. This condition results in the mitochondrial damage, which leads to release of cytochrome c, caspase activation, and then apoptotic cell death [10,11]. Formation of caspases activation resulting in the mediation and amplification of the death signal and eventually in the execution of cell death with all the morphological and biochemical features usually observed [12].

Brain development begins in intrauterine period and continues until the end of the adolescence in mammalians [13]. Brain tissue is rich in polyunsaturated fatty acids which are highly susceptible to oxidative reactions [14,15]. The content of polyunsaturated fatty acids and the degree of unsaturation increase during brain development [16]. The increase in degree of unsaturation of fatty acids increases their susceptibility to peroxidation. The peroxidation is deleterious effects on membrane lipids and membrane functions [14,17]. Thus, our lab is currently exploring the molecular and cellular mechanisms responsible for neuronal apoptosis in developing brain, with a particular focus on the role of ROS. ROS is produced for normal central nervous system development and function. Antioxidant defense systems are eliminate the ROS. However, imbalances, either excess or deficiency, can result in neuronal apoptosis. Developing brain is much more sensitive to environmental exposures

[13]. Many environmental factors can affect developing brain apoptosis. Some of them triggered, others prevented.

**Table 1. Effects of maternal deprivation on SOD and GPx enzymes activities and TBARS levels in the rat brain.**

| | Hippocampus | | |
|---|---|---|---|
| | SOD U/mg protein | GPx U/mg protein | TBARS nmol/mg protein |
| C6 | 0,96±0,06 | 0,0035±0,0003 | 1,61±0,08 |
| MD6 | 1,04±0,11 | 0,0049±0,0001* | 1,41±0,11* |
| C20 | 1,83±0,06** | 0,0044±0,0003** | 1,05±0,06** |
| MD20 | 2,54±0,11* | 0,0037±0,0002* | 1,41±0,08* |

C6, non-deprived and sacrificed in seventh day group; MD6, maternal deprived at postnatal sixth day and then sacrificed in seventh day; C20, non-deprived and sacrificed in twenty first day group; MD20, maternal deprived at postnatal sixth day and then sacrificed in twenty first day group. Data are presented as the mean±SEM. * $P<0.05$ compared with the control of same day. ** $P<0.05$ compared with the group C6. ). (Table is printed in Neurosci Lett. 2005 Aug 12-19; 384(1-2):98-101, with permission.)

## 1. Effects of Maternal Deprivation on Apoptosis in Developing Brain

In most of the mammalians, mother–infant relation is essential for a normal development and the early loss of maternal care may affect the vulnerability of infants over a lifespan [13]. Lack of maternal care may increase the susceptibility to neuropsychiatric disorders such as anxiety, personality disorders, schizophrenia and depression [18]. Several pathogenetic mechanisms, including the disruption in the developmental pattern of the hypothalamic-pituitary-adrenal axis, over-expression of glucocorticoid receptor (GR) gene, and oxidative stress was described for neurobehavioral changes induced by maternal deprivation [19,20,21].

Approximately on postnatal day 4, the infant rat enters SHRP [13,22,23]. This period lasts until postnatal day 14 in rats and is characterized by reduction of adrenal sensitivity of adrenocorticotropic hormone [13]. It appears that low levels of corticosterone during SHRP are essential for a normal development, because of high levels of corticosterone has widespread deleterious effects on the developing brain [23]. So the physiological and behavioral responses to the maternal deprivation may differ in SHRP and out of that period [13,22,23]. This hypo-responsiveness is time- and stressor-specific, as some more severe stressors have been shown to induce a stress response [24]. Recently, we showed that, maternal deprivation in SHRP was shown to protected oxidative stress and apoptosis in infant rat brain. But, on the contrary this result, maternal deprivation in non-SHRP reduced antioxidant enzyme activities and increased lipid peroxidation in hippocampus (table 1) [25]. Anatomical evidence suggests that the hippocampus can be roughly divided into four subfields: dentate gyrus, CA3, CA2, and CA1 [26]. Also on postnatal day 19, the number of neurons was significantly less in hippocampal CA1, CA3 and gyrus dentate after maternal deprivation compared with the control non-deprived group. There was no significant difference in CA2 neuron number among all groups (Figure 1). There are too much TUNEL-positive cells in the maternal deprived group hippocampus at postnatal 19 days old (Figure 2) [27].

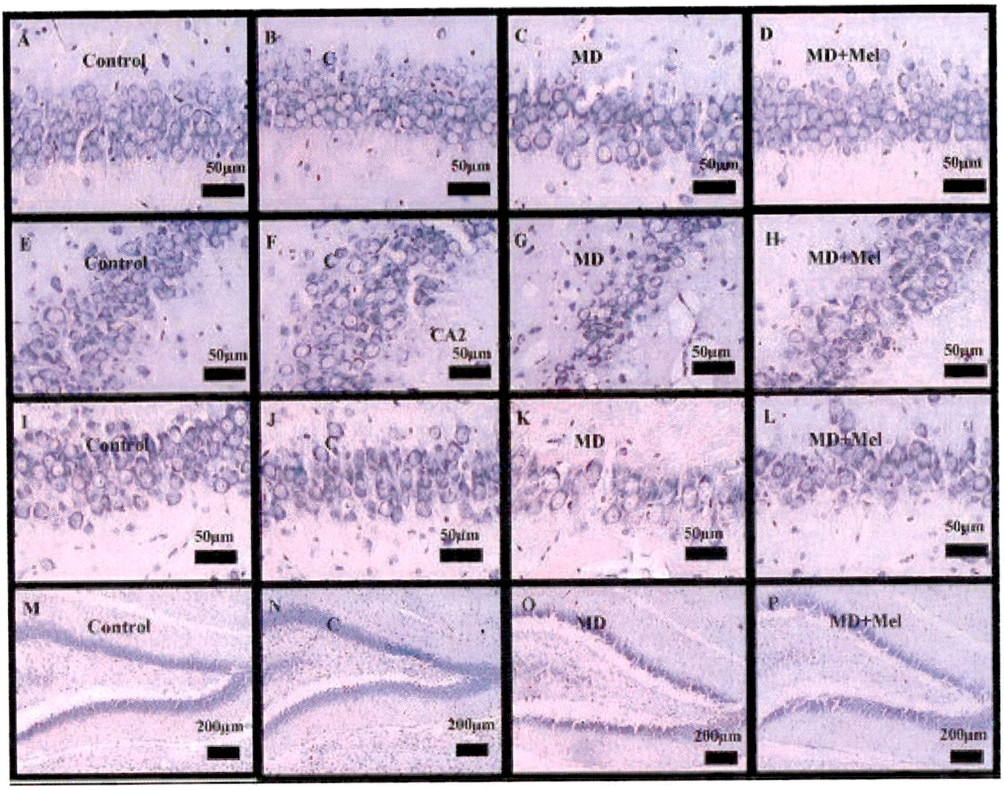


Figure 1. Representative photomicrographs of cresyl-violet stained sections in the hippocampal CA1 (A–D), CA2 (E–H), CA3 (I–L) and gyrus dentate (M–P) 24h after maternal deprivation in 19-day-old rats. When compared with vehicle-treated MD group, the neurons in the regions of CA1, CA3 and gyrus dentate appear well preserved in melatonin-treated MD groups. ). (figure is printed in Neurosci Lett. 2006 May 1;398(1-2):145-50, with permission.).

ROS, namely superoxide and hydroxyl free radicals and hydrogen peroxide, are produced as a consequence of the physiological metabolic reactions and functioning of the central nervous system [28]. Dopamine is metabolized by monoamine oxidase; and hydrogen peroxide is produced as a product [28]. Maternal deprivation increases dopamine turnover in mesolimbic region of brain in laboratory animals [29]. Increased turnover of dopamine evokes an oxidative stress derived from increased production of hydrogen peroxide [28]. Maternal deprivation can also enhances the sensitivity of adrenals to adrenocorticotropin hormone (ACTH) and/or plasma ACTH levels, which then increases plasma corticosterone levels [19,20,21]. Corticosterone increases the toxicity of oxygen radical generators, and may increase the basal level of ROS in the hippocampus and cortex [23,30]. Enhancement of dopamine metabolism may cause oxidative stress in rat brain [31,32]. Generation of ROS can be a major component of decreased cell function and eventual death [28]. The SHRP, which is characterized by low-basal levels of stress hormones and a relative non-responsiveness to external stressors [33,34], was proposed to protect the infant brain against the deteriorating effects of high levels of stress hormones [23]. In our previous study, maternal deprivation in SHRP reduced lipid peroxidation and increased antioxidant enzyme activities in infant rat brain. However, by the termination of SHRP, maternal deprivation increased lipid peroxidation and reduced enzyme activities [25]. When the tissues are exposed to oxidative stress they increase the activity and expression of antioxidant enzymes as a compensatory mechanism against free radical-mediated damage. These results suggest that maternal deprivation-induced neuronal apoptosis is associated with oxidative stress.

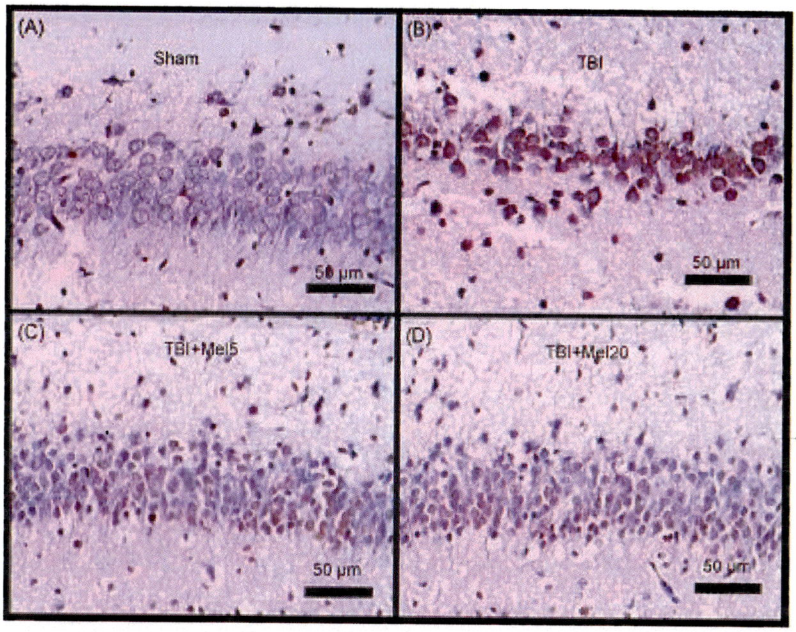


Figure 2. Comparison of terminal transferase dUTP nick end-labeling (TUNEL)-stained sections in the hippocampal CA1 subfield. The number of TUNEL-labeled neurons was reduced in melatonin-treated TBI groups (C and D) compared with vehicle-treated TBI group (B). (figure is printed in Neurosci Lett. 2006 May 1;398(1-2):145-50, with permission.).

## 2. Head Trauma Affect Apoptosis in Developing Brain

Head trauma is a leading cause of death and disability in childhood [35,36]. Although children have better survival rates as compared with adults with head trauma, the long-term consequences are often more pernicious in children due to their age [37]. Traumatic brain injury (TBI) can be classified as primary, which occurs immediately after trauma, and secondary, which may appear several hours or even days later [38]. Two pathogenetic mechanisms have been described for secondary brain injury: excitotoxicity and apoptosis. Head trauma has been demonstrated to activate physiological apoptotic mechanisms in the developing brains [39]. The immature brain is exceedingly vulnerable to apoptotic neurodegeneration triggered by head trauma [40]. This developmental vulnerability period coincides with the brain growth spurt period, which occurs in the first two postnatal weeks in the rats and in the first 2 years old in humans [41]. The severity of the trauma-induced apoptotic neurodegeneration in the brains of 3- to 30-day-old rats had been demonstrated to be age dependent and highest in 7-day-old animals. Apoptotic cell death had reached a peak at 24 h and subsided at 5 days after the traumatic injury [42]. Thus, apoptotic neurodegeneration has been suggested to contribute in an age-dependent effects to neuropathological outcome of head trauma. The neurons and glia may be most vulnerable to apoptosis. And also, the differences in the degree of myelinization and brain water content will allow traumatic forces to transmit easier to deeper brain structures in the immature brain. For these reasons, developing brain is vulnerable to apoptotic neurodegeneration [39,40,43]. Neuronal apoptosis, plays a crucial role in the pathogenesis of traumatic damage in developing brains, which is involved in the delayed and progressive degenerative response to trauma.

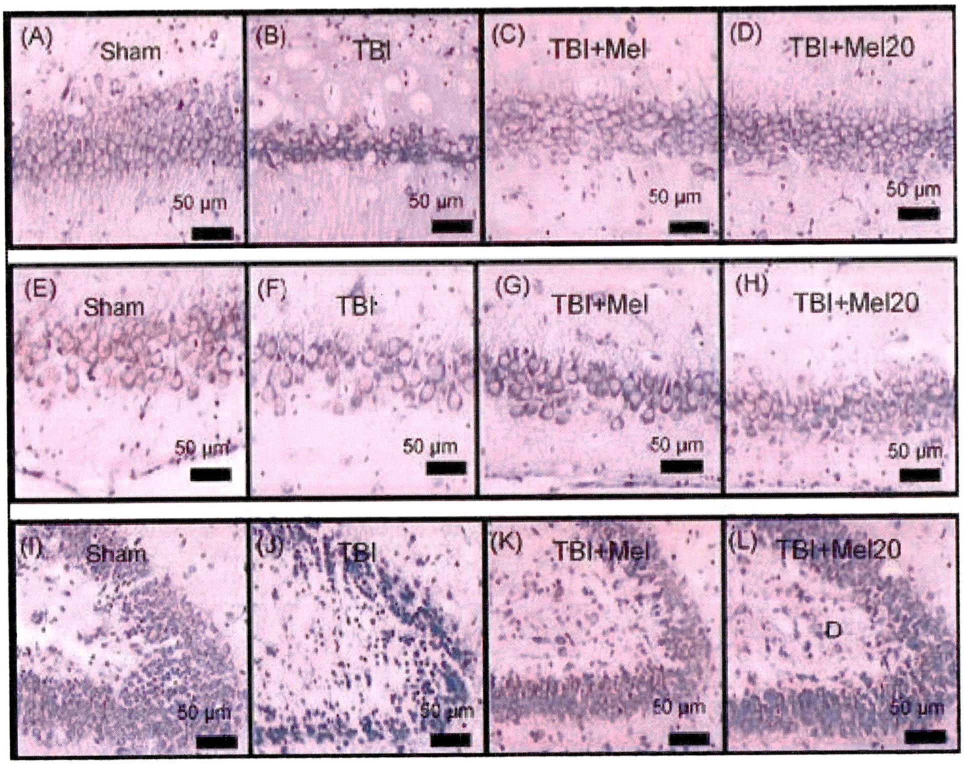


Figure 3. Representative photomicrographs of cresyl-violet stained sections in the hippocampal CA1 (A–D), CA3 (E–H) and gyrus dentate (I–L) 24 h after trauma or sham surgery.When compared with vehicle-treated TBI group, the neurons in the regions examined appear well preserved in melatonin-treated TBI groups. (figure is printed in Neurosci Lett. 2005 Sep 16;385(3):234-9, with permission.).

Two brain structures prominently affected in head trauma are the cortex and the hippocampus [44]. Hippocampal neurons normally play a crucial role in the processing of spatial memory and learning [45,46]. Memory function, in particular, is at risk following head trauma due to the selective vulnerability of the hippocampus. Hippocampal neuron loss is often observed in head trauma and has been recognized to be associated with memory deficits following head trauma [47,48,49]. An important consideration of hippocampal damage following head trauma is the unique pattern of neuron death [50]. Whereas all subfields of the hippocampus have been shown to be damaged following trauma, the most severely injured regions include the CA3 and dentate gyrus [48,49,50]. In our study, we showed that neuronal loss occurred in hippocampal CA1/CA3 and gyrus dentate regions of 7-day-old rats exposed to contusion injury (Fig 3). Interestingly, the contra-lateral hippocampal CA1 and gyrus dentate regions of injured animals also showed significant neuronal loss compared to those of both sham and controls in our study. The number of TUNEL-positive cells in the hippocampal regions was significantly increased following traumatic insult, indicating that neuronal apoptosis induced by head trauma in the immature rat brains (Figure 4) [51]. The cause of the neuronal loss in the contra-lateral brain regions has been suggested to be due to a contra-coup contusive effect and/or delayed neuronal degeneration [44].

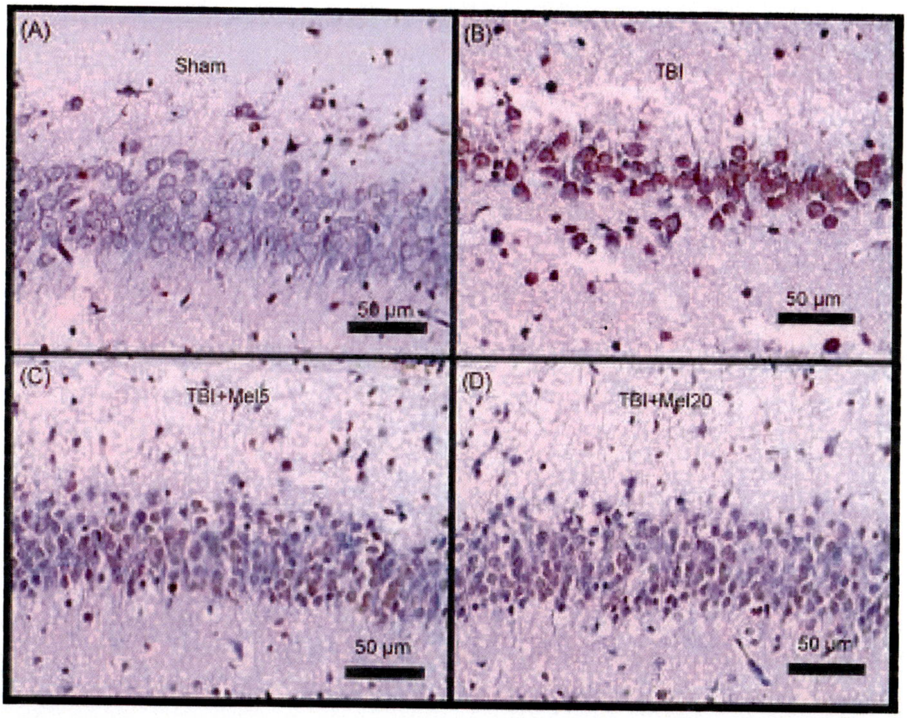


Figure 4. Comparison of terminal transferase dUTP nick end-labeling (TUNEL)-stained sections in the hippocampal CA1 subfield. The number of TUNEL-labeled neurons was reduced in melatonin-treated TBI groups (C and D) compared with vehicle-treated TBI group (B). (figure is printed in Neurosci Lett. 2005 Sep 16;385(3):234-9, with permission.).

The neuronal injury in the cortex and hippocampus has been attributed to direct mechanical damage, excitotoxicity, and oxidative injury after traumatic brain injury [44]. Trauma-induced production of ROS causes cellular injury via several mechanisms, including peroxidation of membrane lipids as well as oxidative damage of proteins and DNA [38]. We also demonstrated that TBI significantly increased lipid peroxidation in both ipsi- and contra-lateral sides of the immature rat brain, there was no compensatory increase in the antioxidant enzymes such as superoxide dismutase (SOD) and glutathione peroxidase (GPx) 24 hours after TBI in 7-day-old rats (Figure 5) [52]. These results indicate that TBI-induced neuronal apoptosis seems related with oxidative stress.

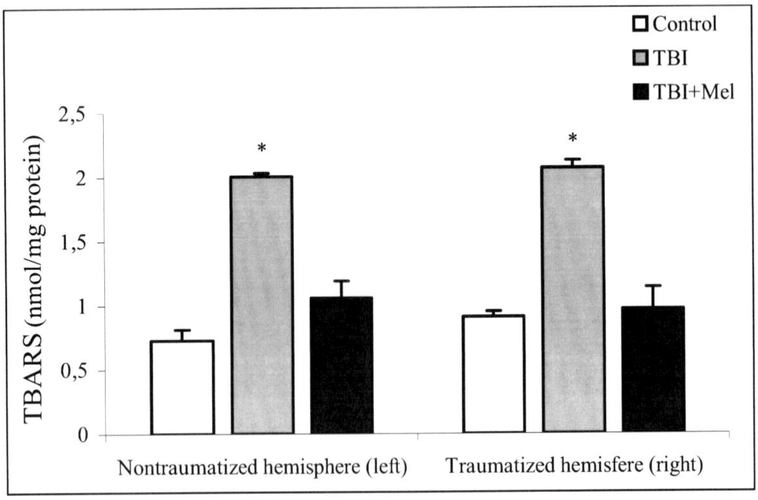


Figure 5. Effects of traumatic brain injury (TBI) on thiobarbituric acid reactive substance (TBARS) levels. TBI increased TBARS levels in the non-traumatized and traumatized hemispheres of the brain. Melatonin prevented the increase in TBARS levels of both brain sides. (figure is printed in Physiol Res. 2005;54(6):631-7, with permission.).

## 3. Ethanol Administration Affects Apoptosis in Developing Brain

Ethanol is one of the most common and important substances that is toxic to the developing fetal brain. Its consumption during pregnancy can cause a disorder termed fetal alcohol syndrome (FAS) [53]. The most disabling feature of the FAS is the dysmorphogenic neuropathological syndrome that includes craniofacial malformations and reduced brain mass, which is associated with a variety of neurobehavioral disturbances, ranging from hyperactivity and attention deficit disorder and learning disabilities in childhood, to major depressive and psychotic disorders in adulthood [53,54,55]. Ethanol, by interfering with major neurotransmitter systems (glutamate and γ-aminobutyric acid [GABA]) during a critical period of development, can cause developing nerve cells to die by apoptosis. Ethanol's apoptogenic effects on the brain coincides with the period of synaptogenesis, also known as the brain growth-spurt period, and it can cause widespread apoptotic neurodegeneration in developing rodent brain [56,57]. Acute ethanol administration produces lipid peroxidation in the brain [58,59,60]. Ethanol is metabolized into cytotoxic acetaldehyde by alcohol dehydrogenase enzyme in the liver. The acetaldehyde must be further oxidized to acetate by acetaldehyde dehydrogenase enzyme, which is present in the brain and is capable of producing ROS [61]. It has been demonstrated that ethanol induces the synthesis of cytochrome P450 (CYP2E1) in the brain. CYP2E1 is present in various brain regions, and it may be an important source of ethanol induced oxidative stress [60]. It also increases the

NADH/ NAD ratio, which causes reduction of ferric iron to ferrous iron, a potent generator of the hydroxyl radical, which can then cause lipid peroxidation [59].

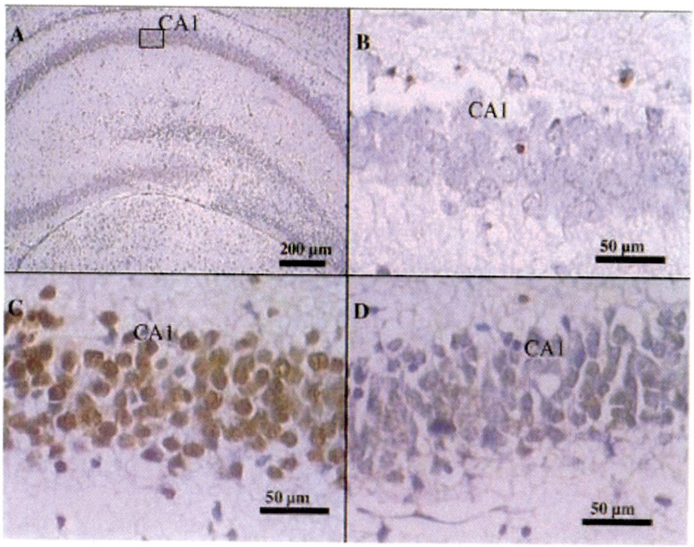


Figure 6. Effect of erythropoietin treatment on the TUNEL immunoreactivity in hippocampus. Panel A shows low magnification view of TUNEL positive cells from CA1 region of mice brain. When compared with control group (B) the number of TUNEL + neurons (dark) in hippocampus increased on ethanol + saline treated mice (C). Erythropoietin (D) significantly reduced number of TUNEL + neurons of region of hippocampus as compared with ethanol + saline treated group. (figure is printed in Brain Res Dev Brain Res. 2005 Dec 7;160(2):146-56, with permission.).

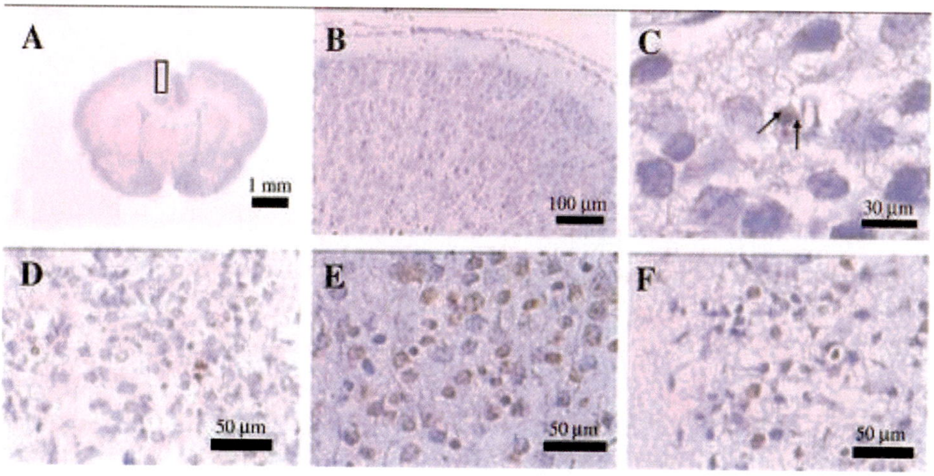


Figure 7. Effect of erythropoietin treatment on the TUNEL immunoreactivity in prefrontal cortex. Lower (A) and higher magnifications (B) of coronal section from prefrontal cortex of mice brain tissues from control group, stained with cresyl violet. Panel C represents nuclear fragmentation which is characteristic for apoptosis from control group. TUNEL positive cells were significantly enhanced with ethanol (E) in ethanol + saline treated group when compared with control (D) group. Erythropoietin (F) significantly reduced number of TUNEL + neurons of region of prefrontal cortex of mice brain tissues as compared with ethanol + saline treated group. (figure is printed in Brain Res Dev Brain Res. 2005 Dec 7;160(2):146-56, with permission.).

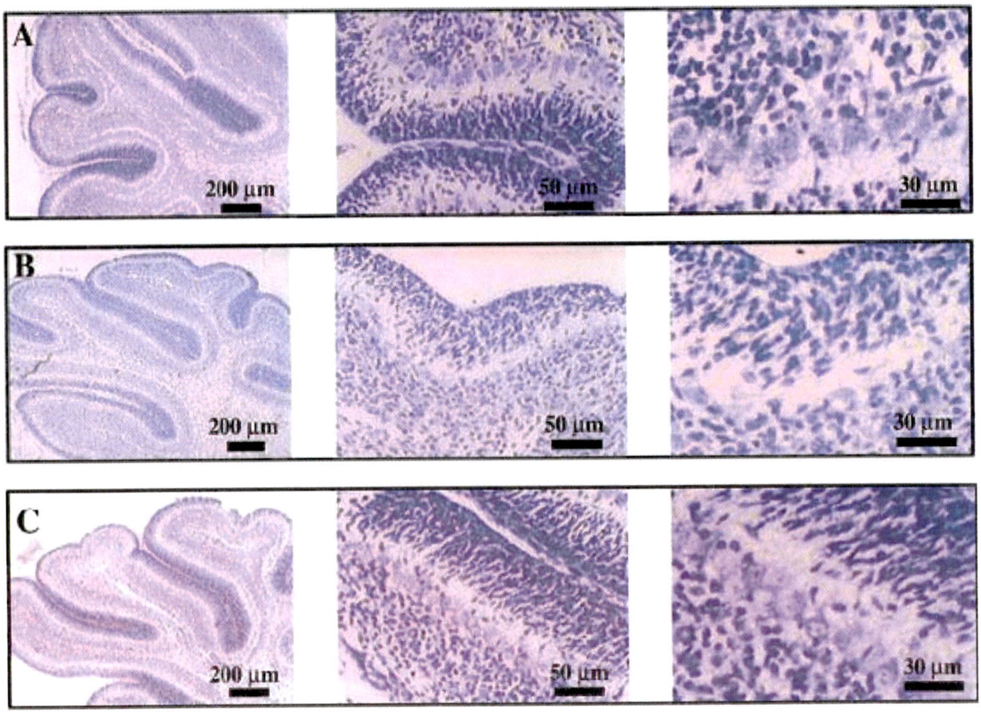


Figure 8. Effect of erythropoietin treatment on the neuron density in cerebellum. Representative pictures obtained by cresyl-violet staining. When compared with control group (A) the number of purkinje neurons of cerebellum decreased on saline treated mice (B). Erythropoietin (C) significantly preserved number of neurons of cerebellum as compared with ethanol + saline treated group. (figure is printed in Brain Res Dev Brain Res. 2005 Dec 7;160(2):146-56, with permission.).

In our study, we demonstrated that administration of ethanol during the early postnatal period is able to induce widespread neurodegeneration in the hippocampus, cerebellum, and prefrontal cortex. We investigated DNA fragmentation in the ethanol-induced brain damage using TUNEL staining. Only few cells with fragmented DNA labeled by TUNEL assay were found in the hippocampus and prefrontal cortex, and cerebellar Purkinje cells of the control mouse. Twenty-four hours after ethanol administration, an increase in the number of TUNEL-positive cells was observed (Figure 6, 7, 8). In addition, we have shown that ethanol caused an increase in TBARS levels, reduced GPx activities, and did not change SOD activities in the mouse brain when compared to the control group (Table 2) [62]. The pharmacological mechanisms of ethanol's effect may result in the production of oxidative stress [58,63]. It has been suggested that ethanol may cause tissue damage through lipid peroxidation [58,64]. Ethanol can enhance reactive oxygen species formation through induction of cytochrome P450 (CYP2E1), which is widely distributed in the brain [64]. The ROS formed from these systems are a source of oxidative stress in the brain [64]. ROS seem to play an important role in ethanol-induced neurotoxicity.

**Table 2. The effects EPO treatment on the activities of GPx and SOD, and TBARS levels in the mouse brain.**

| | SOD<br>U/mg protein | GPx<br>U/mg protein | TBARS<br>nmol/mg protein |
|---|---|---|---|
| Control | 1,570±0,132 | 0,011±0,001 | 0,532±0,043 |
| EtOH + Saline | 1,666±0,062 | 0,010±0,001 | 0,749±0,053* |
| EtOH + EPO | 1,574±0,135 | 0,013±0,001* ** | 0,358±0,020** |

Values are shown as means ± SEM of 8 animals per group. * $P < 0.001$ compared with control group. ** $P < 0.001$ compared with ethanol group. (figure is printed in Brain Res Dev Brain Res. 2005 Dec 7; 160(2):146-56, with permission.)

## 4. Effects of Exercise on Apoptosis in Developing Brain

Regular physical exercise may promote neural health and function. Physical exercise such as wheel running and treadmill running has been shown to increase cell proliferation and survival in the hippocampal dentate gyrus of adult rodents [65,66,67]. We showed that light regular exercise significantly increased the CA1, CA3 neurons numbers in hippocampus and gyrus dentatus regions of young rat brain which continued developing (Fig 9) [68]. Also, no statistically significant difference was observed in the CA2 neuron numbers of hippocampus in our study. The CA2 hippocampal region is relatively resistant to external exposures [69]. The mechanisms of control of neuron number may involve both regulation of cell proliferation or cell death. It has been suggested that there is a relationship between treadmill exercise and apoptosis. It adopted that apoptotic cell death induced by exercise in very metabolically active tissues, such as the heart and skeletal muscle, may be a normal process to remove partially damaged cells in adults. Exercise-induced apoptosis is known to occur in skeletal muscle [70]. Increase in glucocorticoid secretion, intracellular calcium concentration, DNA damage, hypoxia and reactive oxygen species production has been shown to occur during strenuous exercise and these effects may induce apoptosis [70].

Kim et al. [65] have showed that both the easy and moderate exercise do not changes apoptosis in the dentate gyrus of adult rats. In our study, no statistically significant difference was observed in the number of TUNEL-positive cells between the control and exercise groups [68]. This result indicates that light-exercise with fixed intensity does not alter the level of apoptosis in the developing hippocampus. Thus, the light exercise might be insufficient to stimulate apoptosis, and it is possible that animals might have adapted to exercise in our study. Following the adaptation period to training, perceived exercise intensity for rats may be lighter through the exercise period. Thus, the no significant difference for apoptosis may result from reduced relative exercise intensity [68]. Therefore, additional researches focusing on exercise intensity is needed. Increased neuron number may result from both an increase of cell proliferation or a decrease of cell death. The our experiment showed that running activity significantly increases total number of neuron, and did not changed apoptosis of young animals [68]. This result indicates that increased neuron number may be resulted from cell proliferation. Thus, light exercise might be beneficial to developing hippocampal function because it increases the potential for neurogenesis. Exercise modifies developing brain, but the mechanism by which it does so is unknown. Further studies are needed to explain these mechanisms.

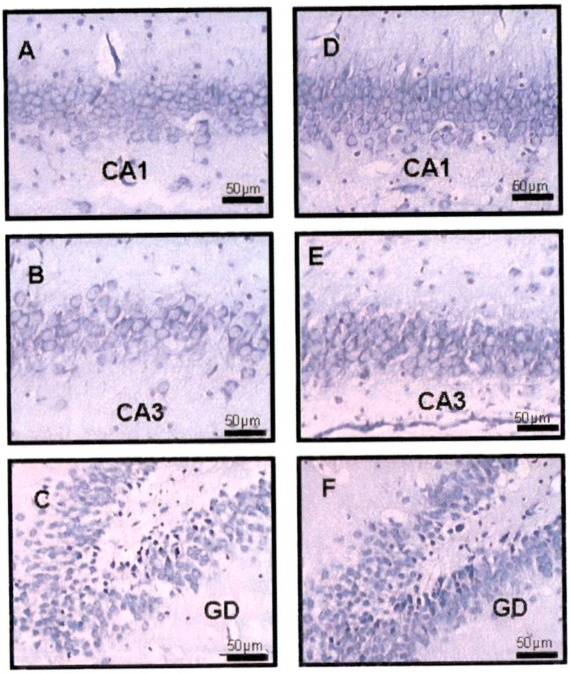


Figure 9. Effect of exercise regimen on the CA1, CA2 and CA3 neuron density in hippocampus and gyrus dentatus. Representative pictures obtained by cresyl-violet staining illustrate the improving effect of exercise on these neurons. Higher magnifications are shown in (B) and (D). Bar=20 µm. (figure is printed in Neurosci Lett. 2005 Aug 5;383(3):241-5, with permission.).

## 5. Effects of the Neuroprotective Agents on Environmental Exposures-Induced Apoptosis in Developing Brains

### *a. Melatonin*

The pineal hormone, melatonin, is a potent free radical scavenger and an antioxidant agent [71]. Melatonin was suggested to reduce the affinity of GR in rat brain and affected on regulation of GR gene expression [72]. Exogenous melatonin decreases the expression of GR and increases cell proliferation in the dentate gyrus of maternally deprivated rats [73]. Melatonin is also associated with the control of the circadian rhythms and the regulation of HPA [74]. Recently, melatonin administration during maternal separation for 7 days was reported to increase cell proliferation in the gyrus dentate of maternally deprivated rats [73]. We demonstrated that melatonin prevents maternal deprivation-induced hippocampal neuron loss in the infant rats at post-SHRP (Figure 1). Furthermore, our data using the TUNEL staining in post-SHRP rats demonstrated that only few cells with fragmented DNA labeled by TUNEL assay were found in the rats treated with melatonin (Figure 2) [27]. The neuroprotective effects of melatonin might be due to its antioxidant and anti-apoptotic properties [71,75]. Moreover, previous studies suggested that melatonin reduced the affinity of GR in rat brain and affected on regulation of GR gene expression [72,76]. Melatonin prevents the hippocampal damage in maternally deprivated rats. Thus, it may be proposed as a beneficial agent to prevent maternal deprivation induced-hippocampal apoptosis.

Melatonin, is also rapidly absorbed of the administration area since its lipophilic nature, and readily passes the blood–brain barrier [71]. In adult rats, melatonin has been shown to reduce contusion volume and to be protective for neurobehavioral dysfunction after experimental head trauma [77,78,79]. But, there are no experimental studies investigating the neuroprotective effects of melatonin against trauma-induced neurodegeneration in immature brains. We demonstrated that melatonin improved spatial memory deficits induced by traumatic injury in the developing rats, which was parallel with neuroprotection against acute hippocampal damage. Indeed, in our study, melatonin was found to prevent the neuronal loss in both ipsi-and contra-lateral regions affected of immature rat brain (Figure 3). There are markedly fewer TUNEL positive cells in the rats treated with melatonin (Figure 4) [51]. We have recently reported the effect of melatonin against oxidative damage induced by TBI in the immature brain [52]. Thus, the neuroprotective effects of melatonin may reflect its role as a free radical scavenger, an antioxidant, an anti-apoptotic agent [52,71,80]. Very recently, melatonin-induced neuroprotection after closed head injury has been suggested to be associated with increased brain antioxidants and attenuated late-phase activation of NF-_B and AP-1 [77]. Previous studies in adult animals have shown that the neuroprotective effect of melatonin is dose-dependent. Mesenge et al. [78] reported that melatonin given as a dose of 1.25 mg/kg 5 min and repeated at 1, 2, and 3 h after head trauma significantly reduced the neurological deficit in mice, which was evaluated at 24 h by a grip test, but neither with the lowest dose (0.625 mg/kg) nor the highest dose (2.5 mg/kg). Beni et al. [77] found that given 1 h after closed head injury in mice, melatonin facilitated recovery during at least 1 week and decreased lesion size, and neuroprotection was achieved with 5 but not 1 or 10 mg/kg. There are several differences between our study and the previous studies for model used, age, the dose and time of melatonin treatment, the type and time of assay after trauma. We investigated whether melatonin given immediately following contusion injury in 7-day-old rat pups would reduce hippocampal damage at 24 h and improve spatial memory deficits 3 weeks after injury. There was no significant difference between the doses of 5 or 20 mg/kg of melatonin in terms of preventing both acute hippocampal damage and long-term functional deficits. A single administration of melatonin immediately after traumatic insult prevents both hippocampal neuron loss and cognitive dysfunction in the developing brains [51]. Thus, it may be proposed as a highly promising agent for preventing the unfavorable outcomes of traumatic brain damage in young children.

### *b. Erythropoietin*

Erythropoietin (EPO) is a hematopoietic cytokine, which has recently been shown to be expressed in the brain [81,82]. EPO is required for normal brain development [83]. It has been shown that EPO has neuroprotective effects in adult rodents [84,85,86]. Although the exact mechanisms of the neuroprotective action of EPO are not completely known, EPO has a role in cell survival signaling cascades [84,86], upregulation of the expression of antiapoptotic proteins [87], modulation of intracellular calcium metabolism [88], inhibition of glutamate release [89] and antiapoptotic [85,86] and antioxidative [90]. We have recently reported beneficial effects of EPO in the neonatal rat brain injury and minimize long-lasting functional deficits throughout brain maturation and into adulthood [91]. The single dose EPO is protected CA1, CA2 region of hippocampus, and prefrontal cortex, and cerebellum, from the ethanol-induced injury in our other study [62]. In the EPO-treated mouse, apoptotic cell counts significantly decreased compared to the saline-treated mouse (Figure 6, Figure 7, Figure 8) [62]. EPO reduces glutamate-induced cytotoxicity and inhibits glutamate release in hippocampal cultures [89,92], regulating apoptosis-related genes and protein expressions [93]. Recent studies have shown that EPO improves abnormalities in the free radical system, reduce free radicals and supports antioxidative defense mechanisms [90,94]. These effects may be responsible for the neuroprotective effect of EPO. In our study, EPO was found to decrease lipid peroxidation, in neonatal mouse brain exposed to the ethanol-induced injury, and also increases GPx activity in the neonatal mouse brain (table 2) [62]. In our study EPO appeared to have a neuroprotective effect against the ethanol-induced neurodegeneration in developing brain [62]. EPO may be potentially beneficial in treating the alcohol-induced brain damage in the perinatal period.

### *c. MK-801*

Glutamate is a prominent excitatory transmitter found in the central nervous system [95], and three primary subtypes of the ion gating glutamate receptors have been identified: N-methyl-D-aspartate (NMDA), α-amino-3-hydro-5-methyl-4-isoxazolpropionic acid (AMPA), and kainate receptors [96]. Excessive glutamate release and subsequent over-activation of specific receptors, especially the NMDA receptor subtype, has been suggested to play a critical role in the pathogenesis of the immature brain to ischemic damage [97,98] and traumatic brain injury [99] as well as that associated with a number of neurodegenerative disorders in adults, including Alzheimer disease and Huntington disease [100]. Hypoxia-ischemia induced excitotoxicity resulting from exposure to excess glutamate via NMDA receptors has been associated with apoptosis in developing brain [98]. Glutamate neurotoxicity occurs after overstimulation of glutamate receptors is associated with a calcium-mediated suppression of membrane protein kinase C activity in developing brain [101].

MK-801, noncompetitive NMDA receptor selective antagonist, has both pro-apoptotic and neuroprotective actions [102]. Several studies in cortical cell cultures from rodents report that the administration of MK-801 alone induces apoptosis by activating caspase-3 [103,104]. On the other hand, blockade of cellular injury induced by glutamate or NMDA by MK-801 may reduce apoptosis [105,106]. As previously reported, treatment with MK-801 has a protective effect against a protein kinase inhibitor that typically induces caspase-dependent cell death [107]. This protection may be due to blocking of glutamate released from damaged cells. MK-801 also prevented hypoxia-induced apoptosis in 7-day-old rat hippocampus, a brain area rich in glutamate receptors [108].

## Conclusion

Programmed cell death has received phenomenal attention in the past few years. Neuronal apoptosis is the common denominator in maternal deprivation, head trauma, intake of ethanol in developing brain. The mechanisms that lead to apoptosis under these different conditions are clearly complex and varied. It is now known that mitochondria play a central regulatory role in apoptosis, particularly through the cytochrome c pathway. Also, mitochondria and radical species are intimately involved in the programmed cell death. Intracellular changes such as increased oxidative stress, depleted glutathione, and decreased reducing equivalents like NADP and NADPH are sufficient to induce the formation of mitochondrial permeability transition pores, leading to the subsequent release of cytochrome c and the activation of the caspase cascade. Uncontrolled and unnecessary apoptosis may occur during developmental period of brain resulting in various pathologies. The neuroprotective agents such as melatonin, erythropoietin may prevent environmental effects induced apoptosis in developing brain. Further experimental studies will need to be performed to confirm such effects of these agents. Clinical studies might be considered following these.

## References


[1] Kerr, J.F.; Wyllie, A.H. & Currie, A.R. (1972). Apoptosis: a basic biological phenomenon with wide-ranging implications in tissue kinetics. *Br. J. Cancer, 26,* 239–257.

[2] Woodle, E.S. & Kulkarni, S. (1998). Programmed cell death. *Transplantation, 66,* 681–691.

[3] Sastry, P.S. & Rao, K.S. (2000). Apoptosis and the nervous system. *J Neurochem, 74,* 1-20.

[4] Bratton, S.B.; MacFarlane, M.; Cain, K. & Cohen, G.M. (2000). Protein complexes activate distinct caspase cascades in death receptor and stress-induced apoptosis. *Exp Cell Res, 256,* 27-33.

[5] Leist, M. & Jaattela, M. (2001). Four deaths and a funeral: from caspases to alternative mechanisms. *Nat. Rev. Mol. Cell Biol, 2,* 589-98.

[6] Salvesen, G.S. & Renatus, M. (2002). Apoptosome: the seven-spoked death machine. *Dev Cell, 2,* 256-7.

[7] Curtin, J.F.; Donovan, M. & Cotter, T.G. (2002). Regulation and measurement of oxidative stress in apoptosis. *J Immunol Methods, 265,* 49-72.

[8] Buttke, T.M. & Sandstrom, P.A. (1994). Oxidative stress as a mediator of apoptosis. *Immunol Today, 15,* 7-10.

[9] Forrest, V.J.; Kang, Y.H.; McClain, D.E.; Robinson, D.H. & Ramakrishnan, N. (1994). Oxidative stress-induced apoptosis prevented by Trolox. *Free Radic Biol Med, 16,* 675-684.

[10] Leon, J.; Acuna-Castroviejo, D.; Escames, G.; Tan, D.X. & Reiter, R.J. (2005). Melatonin mitigates mitochondrial malfunction. *J Pineal Res, 38,* 1-9.

[11] Leon, J.; Acuna-Castroviejo, D.; Sainz, R.M.; Mayo, J.C.; Tan, D.X. & Reiter, R.J. (2004). Melatonin and mitochondrial function. *Life Sci, 75,* 765-790.

[12] Earnshaw, W.C.; Martins, L.M. & Kaufmann, S.H. (1999). Mammalian caspases: structure, activation, substrates, and functions during apoptosis. *Annu Rev Biochem, 68,* 383-424.

[13] Rice, D. & Barone, S.Jr. (2000). Critical periods of vulnerability for the developing nervous system: evidence from humans and animal models. *Environ. Health Perspect, 108,* 511–533.

[14] Halliwell, B. & Chirico, S. (1993). Lipid peroxidation: Its mechanism, measurement, and significance. *Am. J. Clin. Nutr. 57,* 715S–725S.

[15] Sun, G.Y. & Sun, A.Y. (1974). Synaptosomal plasma membranes: Acyl group composition of phosphoclycerides and (Na+ plus K+)-ATPase activity during fatty acid deficiency. *J. Neurochem. 22,* 15–18.

[16] Crawford, M.A. & Sinclair, A.J. (1972). Nutritional influences in evolution of mammalian brain. In: Lipids, malnutrition and the developing brain. Ciba Foundation Symposia. Chichester, NY: *Associated Scientific Publishers;* 267–292.

[17] Mishra, O.P.; Delivoria-Papadopoulos, M.; Cahillane, G. & Wagerle, L.C. (1989). Lipid peroxidation as the mechanism of modification of the affinity of the Na+,K+-ATPase active sites for ATP, K+,Na+ and strophanthidine in vitro. *Neurochem. Res. 14,* 845– 851.

[18] McIntosh, J.; Anisman, H. & Merali, Z. (1999). Short- and long periods of neonatal maternal separation differentially affect anxiety and feeding in adult rats: gender-dependent effects. *Dev. Brain Res. 113,* 97–106.

[19] Ladd, C.O.; Owens, M.J. & Nemeroff, C.B. (1996). Persistent changes in corticotropin-releasing factor neuronal systems induced by maternal deprivation. *Endocrinology, 137,* 1212–1218.

[20] Lee, H.J.; Kim, J.W.; Yim, S.V.; Kim, M.J.; Kim, S.A.; Kim, Y.J.; Kim, C.J. & Chung, J.H. (2001). Fluoxetine enhances cell proliferation and prevents apoptosis in dentate gyrus of maternally separated rats. *Mol. Psychiatry, 6,* 725–728.

[21] Stanton, M.E.; Gutierrez, Y.R. & Levine, S. (1988). Maternal deprivation potentiates pituitary-adrenal stress responses in infant rats, *Behav. Neurosci. 102,* 692–700.

[22] Lehmann, J.; Pryce, C.R.; Bettschen, D. & Feldon, J. (1999). The maternal separation paradigm and adult emotionality and cognition in male and female wistar rats. *Pharm. Biochem. Behav. 64,* 705–715.

[23] Sapolsky, R.M. & Meaney, M.J. (1986). Maturation of the adrenocortical stress response: neuroendocrine control mechanisms and the stress hyporesponsive period. *Brain Res. 39,* 64–76.

[24] Walker, C.D.; Scribner, K.A.; Cascio, C.S. & Dallman, M.F. (2001). The pituitary adrenocortical system of neonatal rats is responsive to stress throughout development in a time-dependent and stressor-specific fashion. *Endocrinology, 128,* 1385–1395.

[25] Uysal, N.; Gonenc, S.; Acikgoz, O.; Pekcetin, C.; Kayatekin, B.M.; Sonmez, A. & Semin, I. (2005). Age-dependent effects of maternal deprivation on oxidative stress in infant rat brain. *Neuroscience Letters, 384,* 98–101.

[26] Amaral, D.G. & Witter, M.P. (1995). Hippocampal formation, in: *The Rat Nervous System,* second ed., Academic, San Diego, 443–493.

[27] Tugyan, K.; Uysal, N.; Ozdemir, D.; Sonmez, U.; Pekcetin, C.; Erbil, G. & Sonmez. A. (2006). Protective effect of melatonin against maternal deprivation-induced acute hippocampal damage in infant rats. *Neuroscience Letters, 398,* 145–150.

[28] Cohen, G. & Spina, M.B. (1989). Deprenyl suppresses the oxidants stress associated with increased dopamine turnover. *Ann. Neurol., 26,* 689-690.

[29] Tamborski, A.; Lucot, J.B. & Hennessey, M.B. (1990). Central dopamine turnover in guinea pig pups during separation from their mothers in a novel environment. *Behav. Neurosci. 104,* 607–611.

[30] McIntosh, J.L.; Hong, E.K. & Sapolsky, R.M. (1998). Glucocorticoids may alter antioxidant enzyme capacity in the brain: baseline studies. *Brain Res. 791,* 209–214.

[31] Gonenc, S.; Acikgoz, O.; Kayatekin, B.M.; Uysal, N. & Akhisaroglu, M. (2000). Effects of footshock stress on superoxide dismutase and glutathione peroxidase enzyme activities and thiobarbituric acid reactive substances levels in the rat prefrontal cortex and striatum. *Neurosci Lett. 289,* 107-10.

[32] Uysal, N.; Acikgoz, O.; Gonenc, S.; Kayatekin, B.M.; Kiray, M. Sonmez, A. & Semin, I. (2005). Effects of acute footshock stress on antioxidant enzyme activities in the adolescent rat brain. *Physiol Res. 54,* 437-42.

[33] Levine, S. (1994). The ontogeny of the hypothalamic-pituitary-adrenal axis. The influence of maternal factors. *Ann. N. Y. Acad. Sci. 746,* 275–288.

[34] Rosenfeld, P.; Wetmore, J.B. & Levine, S. (1992). Effects of repeated maternal separations on the adrenocortical response to stress of preweanling rats. *Physiol. Behav. 52,* 787–791.

[35] Adelson, P.D. & Kochanek, P.M. (1998). Head injury in children. *J. Child Neurol. 13,* 2–15.

[36] Koskiniemi M.; Kyykka, T.; Nybo, T. & Jarho, L. (1995). Long-term outcome after severe brain injury in preschoolers is worse than expected. *Arch. Pediatr. Adolesc. Med. 149,* 249–254.

[37] Guthrie, E.; Mast, J.; Richards, P.; McQuaid, M. & Pavlakis, S. (1999). Traumatic brain injury in children and adolescents. *Child Adolesc. Psychiatr. Clin. N. Am. 8,* 807–826.

[38] Bayir, H.; Kochanek, P.M. & Clark, R.S. (2003). Traumatic brain injury in infants and children: mechanisms of secondary damage and treatment in the intensive care unit. *Crit. Care Clin. 19,* 529–549.

[39] Bittigau, P.; Sifringer, M.; Felderhoff-Mueser, U.; Hansen, H.H. & Ikonomidou, C. (2003). Neuropathological and biochemical features of traumatic injury in the developing brain. *Neurotox. Res. 5,* 475–490.

[40] Bittigau, P.; Sifringer, M.; Felderhoff-Mueser, U. & Ikonomidou, C. (2004). Apoptotic neurodegeneration in the context of traumatic injury to the developing brain. *Exp. Toxicol. Pathol. 56*, 83–89.

[41] Dobbing, J. & Sands, J. (1979). Comparative aspects of the brain growth spurt. *Early Hum. Dev. 3*, 79–83.

[42] Bittigau, P.; Sifringer, M.; Pohl, D.; Stadthaus, D.; Ishimaru, M.; Shimizu, H.; Ikeda, M.; Lang, D.; Speer, A.; Olney, J.M. & Ikonomidou, C. (1999). Apoptotic neurodegeneration following trauma is markedly enhanced in the immature brain. *Ann. Neurol. 45*, 724–735.

[43] Felderhoff-Mueser, U. & Ikonomidou, C. (2000). Mechanisms of neurodegeneration after paediatric brain injury. *Curr. Opin. Neurol. 13,* 141–145.

[44] Tong, W.; Igarashi, T.; Ferriero, D.M. & Noble, L.J. (2002). Traumatic brain injury in the immature mouse brain: characterization of regional vulnerability. *Exp. Neurol. 176*, 105–116.

[45] Morris, R.G.M.; Garrud, P.; Rawlins, J.N.P. & O'Keefe, J. (1982). Place navigation impaired in rats with hippocampus lesions. *Nature 297*, 681–683.

[46] Olton, D.S.; Becker, J.T. & Handelmann, G.E. (1979). Hippocampus space and memory. *Behav. Brain Sci. 2*, 313–365.

[47] Kotapka, M.J.; Gennarelli, T.A.; Graham, D.I.; Adams, J.H.; Thibault, L.E.; Ross, D.T. & Ford, I. (1991). Selective vulnerability of hippocampal neurons in acceleration-induced experimental head injury. *J. Neurotrauma 8*, 247–258.

[48] Lowenstein, D.H.; Thomas, M.J.; Smith, D.H. & McIntosh, T.K. (1992). Selective vulnerability of dentate hilar neurons following traumatic brain injury: A potential mechanistic link between head trauma and disorders of the hippocampus. *J. Neurosci. 12*, 4846–4853.

[49] Smith, D.H.; Lowenstein, D.H.; Gennarelli, T.A. & McIntosh, T.K. (1994). Persistent memory dysfunction is associated with bilateral hippocampal damage following experimental brain injury. *Neurosci. Lett. 168*, 151–154.

[50] Leoni, M.J.; Chen, X.H.; Mueller, A.L.; Cheney, J.; McIntosh, T.K. & Smith, D.H. NPS 1506 attenuates cognitive dysfunction and hippocampal neuron death following brain trauma in the rat. *Exp. Neurol. 166,* 442–449.

[51] Ozdemir, D.; Tugyan, K.; Uysal, N.; Sonmez, U.; Sonmez, A.; Acikgoz, O.; Ozdemir, N.; Duman, M. & Ozkan, H. (2005). Protective effect of melatonin against head trauma-induced hippocampal damage and spatial memory deficits in immature rats. *Neurosci Lett, 385*, 234-9.

[52] Ozdemir, D.; Uysal, N.; Gonenc, S.; Acikgoz, O.; Sonmez, A.; Topcu, A.; Ozdemir, N.; Duman, M.; Semin, I. & Ozkan, H. (2005). Effect of melatonin on brain oxidative damage induced by traumatic brain injury in immature rats. *Physiol. Res. 54*, 631-637.

[53] Clarren, S.K.; Alvord, E.C.; Sumi, S.M.; Streissguth, A.P. & Smith, D.W. (1978). Brain malformations related to prenatal exposure to ethanol. *J. Pediatr. 92*, 64–67.

[54] Famy, C.; Streissguth, A.P. & Unis, A.S. (1998). Mental illness in adults with fetal alcohol syndrome or fetal alcohol effects. *Am. J. Psychiatry 155*, 552– 554.

[55] Kerns, K.A.; Don, A.; Mateer, C.A. & Streissguth, A.P. (1997). Cognitive deficits in nonretarded adults with fetal alcohol syndrome. *J. Learn. Disabil. 30*, 685– 693.

[56] Olney, J.W.; Tenkova, T.; Dikranian, K.; Qin, Y.Q.; Labruyere, J. & Ikonomidou, C. (2002). Ethanol-induced apoptotic neurodegeneration in the developing C57BL/6 mouse brain. *Brain Res. Dev. Brain Res. 133*, 115– 126.

[57] Pierce, D.R. & West, J.R. (1987). Differential deficits in regional brain growth induced by postnatal alcohol. *Neurotoxicol. Teratol. 9*, 129– 141.

[58] Gonenc, S.; Uysal, N.; Acikgoz, O.; Kayatekin, B.M.; Sonmez, A.; Kiray, M.; Aksu, I.; Gulecer, B.; Topcu, A. & Semin, I. (2005). Effects of melatonin on oxidative stress and spatial memory impairment induced by acute ethanol treatment in rats. *Physiol Res. 54*,341-8.

[59] Mansouri, A.; Demeilliers, C.; Amsellem, S.; Pessayre, D. & Fromenty, B. (2001). Acute ethanol administration oxidatively damages and depletes mitochondrial DNA in mouse liver, brain, heart, and skeletal muscles: protective effects of antioxidants. *J. Pharmacol. Exp. Ther. 298*, 737– 743.

[60] Somani, S.M.; Husain, K.; Diaz-Phillips, L.; Lanzotti, D.J.; Kareti, K.R. & Trammell, G.L. (1996). Interaction of exercise and ethanol on antioxidant enzymes in brain regions of the rat. *Alcohol 13*, 603– 610.

[61] Schlorff, E.C.; Husain, K. & Somani, S.M. (1999). Dose- and time-dependent effects of ethanol on plasma antioxidant system in rat. *Alcohol 17,* 97–105.

[62] Kumral, A.; Tugyan, K.; Gonenc, S.; Genc, K.; Genc, S.; Sonmez, U.; Yilmaz, O.; Duman, N.; Uysal, N. & Ozkan, H. (2005). Protective effects of erythropoietin against ethanol-induced apoptotic neurodegenaration and oxidative stress in the developing C57BL/6 mouse brain. *Brain Res Dev Brain Res. 160*, 146-56.

[63] Malbohan, I.; Janebova, M.; Nespor, K. & Sun, G.Y. (2001). Ethanol and oxidative stress. *Alcohol Clin. Exp. Res. 25,* 237S– 243S.

[64] Montoliu, C.; Sancho-Tello, M.; Azorin, I.; Burgal, M.; Valles, S.; Renau-Piqueras, J. & Guerri, C. (1995). Ethanol increases cytochrome P4502E1 and induces oxidative stress in astrocytes. *J. Neurochem. 65,* 2561– 2570.

[65] Kim, S.H.; Kim, H.B.; Jang, M.H.; Lim, B.V.; Kim, Y.J.; Kim, Y.P.; Kim, S.S.; Kim, E.H. & Kim, C.J. (2002). Treadmill exercise increases cell proliferation without altering of apoptosis in dentate gyrus of Sprague–Dawley rats. *Life Sci. 71,* 1331–1340.

[66] Kim, Y.P.; Kim, H.B.; Jang, M.H.; Lim, B.V.; Kim, Y.J.; Kim, H.; Kim, S.S.; Kim, E.H. & Kim, C.J. (2003). Magnitude and time-dependence of the effect of treadmill exercise on cell proliferation in the dentate gyrus of rats. *Int. J. Sports Med. 24,* 108–113.

[67] van Praag, H.; Kempermann, G. & Gage, F.H. (1999). Running increases cell proliferation and neurogenesis in the adult mouse dentate gyrus. *Nat. Neurosci. 2,* 266–270.

[68] Uysal, N.; Tugyan, K.; Kayatekin, B.M.; Acikgoz, O.; Bagriyanik, H.A.; Gonenc, S.; Ozdemir, D.; Aksu, I.; Topcu, A. & Semin, I. (2005). The effects of regular aerobic exercise in adolescent period on hippocampal neuron density, apoptosis and spatial memory. *Neurosci Lett. 383*, 241-5.

[69] Chung, S.Y. & Han, S.H. (2003). Melatonin attenuates kainic acid-induced hippocampal neurodegeneration and oxidative stress through microglial inhibition. *J Pineal Res. 34*, 95–102.

[70] Carraro, U. & Franceschi, C. (1997). Apoptosis of skeletal and cardiac muscles and physical exercise. *Aging-Clin. Exp. Res. 9*, 919–934.

[71] Reiter, R.J. (1998). Oxidative damage in the central nervous system: protection by melatonin. *Prog. Neurobiol. 56*, 359–384.

[72] Marinova, C.; Persengiev, S.; Konakchieva, R.; Ilieva, A. & Patchev, V. (1991). Melatonin effects on glucocorticoid receptors in rat brain and pituitary: significance in adrenocortical regulation. *Int. J. Biochem. 23*, 479–481.

[73] Kim, M.J.; Kim, H.K.; Kim, B.S. & Yim, S.V. (2004). Melatonin increases cell proliferation in the dentate gyrus of maternally separated rats. *J. Pineal Res. 37*, 193–197.

[74] Quay, W.B. (1989). Changes with darkness in regional brain 5-hydroxytryptamine and 5-hydroxyindole acetic acid: local differences with pinealectomy, sham surgery, and melatonin. *Neurochem. Res. 14*, 957–961.

[75] Leon, J.; Acuna-Castroviejo, D.; Escames, G. Tan, D.X. & Reiter, R.J. (2005). Melatonin mitigates mitochondrial malfunction. *J. Pineal Res. 38*, 1–9.

[76] Persengiev, S.P. (1999). Multiple domains of melatonin receptor are involved in the regulation of glucocorticoid receptor-induced gene expression. *J. Steroid Biochem. Mol. Biol. 68*, 181–187.

[77] Beni, S.M.; Kohen, R.; Reiter, R.J.; Tan, D.X. & Shohami, E. (2004). Melatonin induced neuroprotection after closed head injury is associated with increased brain antioxidants and attenuated late-phase activation of NF-_B and AP-1. *FASEB J. 18*, 149–151.

[78] Mesenge, C.; Margaill, I.; Verrecchia, C.; Allix, M.; Boulu, R.G. & Plotkine, M. Protective effect of melatonin in a model of traumatic brain injury in mice. *J. Pineal Res. 25*, 41–46.

[79] Sarrafzadeh, A.S.; Thomale, U.W.; Kroppenstedt, S.N. & Unterberg, A.W. (2000). Neuroprotective effect of melatonin on cortical impact injury in the rat. *Acta Neurochir. 142*, 1293–1299.

[80] Feng, Z.; Chang, Y.; Cheng, Y.; Zhang, B.L.; Qu, Z.W.; Qin, C. & Zhang, J.T. (2004). Melatonin alleviates behavioral deficits associated with apoptosis and cholinergic system dysfunction in the APP 695 transgenic mouse model of Alzheimer's disease. *J. Pineal Res. 37*, 129–136.

[81] Dame, C.; Juul, S.E. & Christensen, R.D. (2001). The biology of erythropoietin in the central nervous system and its neurotrophic and neuroprotective potential. *Biol. Neonate 79*, 228–235.

[82] Siren, A.L. & Ehrenreich, H. (2001). Erythropoietin—A novel concept for neuroprotection. *Eur. Arch. Psychiatry Clin. Neurosci. 251*, 179– 184.

[83] Yu, X.; Shacka, J.J.; Eells, J.B.; Suarez-Quian, C.; Przygodzki, R.M.; Beleslin-Cokic, B.; Lin, C.S.; Nikodem, V.M.; Hempstead, B.; Flanders, K.C.; Costantini, F. & Noguchi, C.T. (2002). Erythropoietin receptor signaling is required for normal brain development. *Development 129*, 505– 516.

[84] Celik, M.; Gokmen, N.; Erbayraktar, S.; Akhisaroglu, M.; Konakci, S.; Ulukus, C.; Genc, S.; Genc, K.; Sagiroglu, E.; Cerami, A. & Brines, M. (2002). Erythropoietin prevents motor neuron apoptosis and neurologic disability in experimental spinal cord ischemic injury. *Proc. Natl. Acad. Sci. U. S. A. 99*, 2258–2263.

[85] Sakanaka, M.; Wen, T.C.; Matsuda, S.; Masuda, S.; Morishita, E.; Nagao, M. & Sasaki, R. (1998). In vivo evidence that erythropoietin protects neurons from ischemic damage. *Proc. Natl. Acad. Sci. U. S. A. 95*, 4635– 4640.

[86] Siren, A.L.; Fratelli, M.; Brines, M.; Goemans, C.; Casagrande, S.; Lewczuk, P.; Keenan, S.; Gleiter, C.; Pasquali, C.; Capobianco, A.; Mennini, T.; Heumann, R.; Cerami, A.; Ehrenreich, H. & Ghezzi, P. (2001). Erythropoietin prevents neuronal apoptosis after cerebral ischemia and metabolic stress. *Proc. Natl. Acad. Sci. U. S. A. 98*, 4044– 4049.

[87] Wen, T.C.; Sadamoto, Y.; Tanaka, J.; Zhu, P.X.; Nakata, K.; Ma, Y.J.; Hata, R. & Sakanaka, M. (2002). rythropoietin protects neurons against chemical hypoxia and cerebral ischemic injury by up-regulating Bcl-xL expression. *J. Neurosci. Res. 67*, 795–803.

[88] Koshimura, K.; Murakami, Y.; Sohmiya, M.; Tanaka, J. & Kato, Y. (1999) Effects of erythropoietin on neuronal activity. *J. Neurochem. 72*, 2565– 2572.

[89] Kawakami, M.; Sekiguchi, M.; Sato, K.; Kozaki, S. & Takahashi, M. (2001). Erythropoietin receptor-mediated inhibition of exocytotic glutamate release confers neuroprotection during chemical ischemia. *J. Biol. Chem. 276*, 39469–39475.

[90] Kumral, A.; Gonenc, S.; Acikgoz, O.; Sonmez, A.; Genc, K.; Yilmaz, O.; Gokmen, N.; Duman, N. & Ozkan, H. (2005). Erythropoietin increases glutathione peroxidase enzyme activity and decreases lipid peroxidation levels in hypoxic– ischemic brain injury in neonatal rats. *Biol. Neonate 87*, 15–18.

[91] Kumral, A.; Uysal, N.; Tugyan, K.; Sonmez, A.; Yilmaz, O.; Gokmen, N.; Kiray, M.; Genc, S.; Duman, N.; Koroglu, T.F.; Ozkan, H. & Genc, K. (2004). Erythropoietin improves long-term spatial memory deficits and brain injury following neonatal hypoxia– ischemia in rats. *Behav. Brain Res. 153*, 77– 86.

[92] Morishita, E.; Masuda, S.; Nagao, M.; Yasuda, Y. & Sasaki, R. (1997). Erythropoietin receptor is expressed in rat hippocampal and cerebral cortical neurons, and erythropoietin prevents in vitro glutamate-induced neuronal death. *Neuroscience 76*, 105– 116.

[93] Wen, T.C.; Sadamoto, Y.; Tanaka, J.; Zhu, P.X.; Nakata, K.; Ma, Y.J.; Hata, R.; Sakanaka, M. (2002). Erythropoietin protects neurons against chemical hypoxia and cerebral ischemic injury by up-regulating Bcl-xL expression. *J. Neurosci. Res. 67*, 795–803.

[94] Solaroglu, I.; Solaroglu, A.; Kaptanoglu, E.; Dede, S.; Haberal, A.; Beskonakli, E. & Kilinc, K. (2003). Erythropoietin prevents ischemia– reperfusion from inducing oxidative damage in fetal rat brain. *Child's Nerv. Syst. 19*, 19– 22.

[95] Cotman, C.W.; Monaghan, D.T.; Ottersen, O.P. & Storm-Mathisen, J. (1987). Anatomical organization of excitatory amino acid receptors and their pathways. *Trends Neurosci 10*, 273–280.

[96] Watkins, J.C.; Krogsgaard-Larsen, P.; Honore, T. (1990). Structure-activity relationships in the development of excitatory amino acid receptor agonists and competitive antagonists. *Trends Pharmacol Sci. 11*, 25-33.

[97] Mishra, O.P.; Fritz, K.I. & Delivoria-Papadopoulos, M. (2001). NMDA receptor and neonatal hypoxic injury. *Ment Retard Dev Disabil Res Rev 7*, 249 –253.

[98] Barks, J.D.E. & Silverstein, F.S. (1992). Excitatory amino acids contribute to the pathogenesis of perinatal hypoxic-ischemic brain injury. *Brain Pathol 2*, 235–243.

[99] Choi, D.W. (1988). Glutamate neurotoxicity and diseases of the nervous system. *Neuron.1*, 623-34.

[100] Meldrum, B. & Garthwaite, J. (1990). Excitatory amino-acid neurotoxicity and neurodegenerative disease. *Trends in pharmacological sciences 11*, 379-387.

[101] Durkin, J.P.; Tremblay, R.; Buchan, A.; Blosser, J.; Chakravarthy, B.; Mealing, G.; Morley, P.; Song, D. (1996). An early loss in membrane protein kinase C activity precedes the excitatory amino acid-induced death of primary cortical neurons. *J Neurochem 66*, 951–962.

[102] Husson, I.; Mesples, B.; Medja, F.; Leroux, P.; Kosofsky, B. & Gressens, P. (2004). Methylphenidate and MK-801, an *N*-methyl-d-aspartate receptor antagonist: shared biological properties. *Neuroscience 125*, 163–170.

[103] Takadera, T.; Matsuda, I. & Ohyashiki, T. (1999). Apoptotic cell death and caspase-3 activation induced by *N*-methyl-d-aspartate receptor antagonists and their prevention by insulin-like growth factor I. *J Neurochem 73*, 548–556.

[104] Yoon, W.J.; Won, S.J.; Ryu, B.R. & Gwag, B.J. (2003) Blockade of ionotropic glutamate receptors produces neuronal apoptosis through the Bax-cytochrome C-caspase pathway: the causative role of Ca2+ deficiency. *J Neurochem 85*, 525–533.

[105] Ankarcrona, M.; Dypbukt, J.M.; Bonfoco, E.; Zhivotovsky, B.; Orrenius, S.; Lipton, S.A. & Nicotera, P. (1995). Glutamate-induced neuronal death: a succession of necrosis or apoptosis depending on mitochondrial function. *Neuron 15*, 961–973

[106] Nicotera, P.; Leist, M. & Manzo, L. (1999). Neuronal cell death: a demise with different shapes. *Trends Pharmacol Sci 20*, 46–51.

[107] Almaas, R.; Pytte, M.; Lindstad, J.K.; Wright, M.; Saugstad, O.D.; Pleasure, D. & Rootwelt, T. (2003). Acidosis has opposite effects on neuronal survival during hypoxia and reoxygenation. *J Neurochem 84*, 1018–1027.

[108] Grojean, S.; Pourie, G.; Vert, P. & Daval, J. (2003). Differential neuronal fates in the CA1 hippocampus after hypoxia in newborn and 7-day-old rats: effects of pretreatment with MK-801. *Hippocampus, 13*, 970-7.

In: Neuronal Cell Apoptosis Research
Editor: Relif J. Shandling, pp. 179-193
ISBN 1-60021-456-8


*Chapter 8*

# APOPTOSIS IN ECOTOXICOLOGY RESEARCH

***André Amaral and Armindo Rodrigues***
Center of Research in Natural Resources (CIRN) and Department of Biology, University of the Azores, Ponta Delgada, Portugal.

## ABSTRACT

Trying to protect ecosystems and humans as well as trying to answer several questions about the effect, transport and fate of natural and synthetic toxicants, ecotoxicology uses tools and methods from different disciplinary areas and resorts to biological markers indicative of exposure to or effects of toxicants. It is known that cellular death can be promoted by several toxicants and pathological conditions. Generally, when the concentration of a toxicant is high enough to cause irreparable cellular injuries or large disorders in the cellular milieu the cell dies by necrosis, through a process without energy consumption, but on the other hand, it is much more common affected cells to die by apoptosis, which consumes energy.

Apoptosis is a highly conserved and regulated mechanism, essential to organ development, homeostasis and function. This mechanism involved in cell turnover accounts for normal and pathological modifications of tissues, it may occur spontaneously in malignant tumors, and its induction or suppression in the beginning of disease can have huge impact on tissue and organ performance. Additionally, apoptosis can be a consequence of the action of several toxicants, inorganic and organic, present in the air, food, and water to which living organisms are exposed. Trace metals, such as cadmium, mercury, copper, and zinc, individually or in mixtures, are known to induce apoptosis when biological systems are exposed within certain concentration limits. The same phenomenon can occur when organisms are exposed to organic compounds, such as polcycyclic aromatic hydrocarbons (PAHs), polychlorinated biphenyls (PCBs), dioxins, and other persistent pollutants. Apoptosis, which is expressed by specific molecular and cellular morphological features, is therefore an essential issue of research in ecotoxicology since it is strongly correlated to the action of many toxicants and can be important as a biomarker of such action. Thus, apoptosis and its use as a biomarker of inorganic and organic toxicants are here addressed in an ecotoxicological approach. The benefit of apoptosis detection in biomonitoring programs is also emphasized in this chapter.

# INTRODUCTION

## Ecotoxicology: the Concept

The word ecotoxicology is recent and it was firstly defined by Truhaut (1977) as the branch of toxicology concerned with the integrated study of toxic effects, caused by natural and synthetic toxicants, over the ecosystem and its biotic component, i.e. animals, including humans, plants and microorganisms. Inherent to this definition is the research of how and to what level is the biota exposed to those toxicants, and consequently involves also the study of the way of release, transport and transformation of chemicals in the environment (Wright and Welbourn, 2002). Nowadays, ecotoxicology is defined as the study of toxicants and their effects on the constituents of the biosphere, including humans (Newman *et al.*, 2002). To reach the main objective of ecotoxicology, i.e. protect ecosystems from the effects of hazardous toxicants (Landis and Yu, 1995; Connell *et al.*, 1999), this science tries to achieve other more specific goals like the evaluation, the monitoring and the prediction of fate of toxicants that are introduced in the environment (Moriarty, 1983; Calow, 1994).

By studying the impacts of toxicants over the structure and function of ecological systems, ecotoxicology resorts to a multidisciplinary approach in which intervene geology, chemistry, biology, epidemiology, toxicology and pharmacology, among others, and which in practice aims at the perception of the interaction between toxicants and environment, since this interaction controls the availability of those chemicals to the biota (Depledge *et al.*, 1994; Connell *et al.*, 1999). Ecotoxicology looks also for the identification of biological molecules that act as targets of toxicants. And lastly, it aspires to elaborate appropriate logarithms that translate the micro-effect of the interaction of toxicants and their specific molecular receptors to the macro-effects of those same toxicants on the environment (Depledge *et al.*, 1994; Landis and Yu, 1995; Connell *et al.*, 1999).

Questions about the effect, transport and fate of certain toxicant are posed at different levels of biological and physical organization (Newman *et al.*, 2002). However, it is necessary to have in mind that the activity of toxicants suffers the influence of several factors, biotic and abiotic, that regulate their absorption, accumulation, transformation and elimination in the organisms. These factors include: (i) physico-chemical properties of toxicants but also of the surrounding media; (ii) time and mode of exposure; (iii) interactions among toxicants; and (iv) biological characteristics of the species or individual (Phillips, 1980; Depledge *et al.*, 1994; Landis and Yu, 1995; Jouany, 1998; Connell *et al.*, 1999; Cunningham *et al.*, 2005).

The fundamental problem in environmental risk assessment consists in making a valid prediction of risk that relates to the release of a toxicant into the environment with potential adverse effects to the health of living organisms. These effects start with the release of a toxicant into the air, water, soil or food. Traditionally, environmental risk is assessed through quantification of chemical residues in samples from the environment, obtained from the affected sites, and importance given only to the physico-chemical properties of toxicants. Although useful it has some disadvantages. First, quantification of chemical residues in environmental matrixes is complex and can have a high cost per sample. Second, bioavailability cannot be measured by this way. To overcome these difficulties, it is common to perform analysis to chemical residues from tissues or organs of the contaminated living organism, but still this is expensive and more time consuming than analyzing environmental matrixes. A method based on biological indicators or markers solves this problem since it gives more emphasis to biological factors that regulate many of the interactions and transformations chemicals undergo, inside and outside living organisms. This approach allows knowing which is the bioavailable fraction of a certain toxicant, what are the routes of entry and what are their targets for action and accumulation. It tries also to understand how a toxicant is transformed by the different metabolic systems and how detoxification occurs, i.e. how the organism neutralizes and impedes a toxicant from reacting with biomolecules. Therefore, the use of biological indicators or markers provides a direct measure of the effects that a toxicant has on the affected species (Forbes and Forbes, 1994; Connell *et al.*, 1999; Cajaraville *et al.*, 2000; Klaasen and Watkins III, 2001).

## Bioindicators, Biomarkers, and Apoptosis

A bioindicator is defined as a measure, or index of measures, or a model of biological responses that characterize an ecosystem or one of its critical components, and may reflect biological, chemical or physical attributes of an ecological condition. The main purpose of a bioindicator is to characterize the state of a certain population, community or ecosystem and predict any significant changes to that state (Walker *et al.*, 2000; EPA, 2005). The use of these indicators allows also to increase the sensitivity of analytical procedures and to summarize a complex signal of pollution (Beeby, 2001). The presence or the absence and the condition of a certain species, the specific richness and the diversity of the ecosystem are well known examples of bioindicators.

Bioindicators are especially used in environmental monitoring programs to evaluate several environmental parameters on a regular temporal basis. One of the pioneer monitoring programs was the Mussel Watch, which in the US in 1975 aimed to measure the influence on living organisms of artificial radionuclides, hydrocarbons, and metals occurring in sea water (O'Connor, 1998). In the same year the program MED POL for the Mediterranean Sea was created with the objective of monitoring pollution by heavy metals, TBT, organohalogenated compounds, and others (Gabrielides, 1995).

To be used in monitoring programs, the chosen species should, if possible, fulfill the following criteria (Landis and Yu, 1995): (i) to be sedentary and abundant in the study area; (ii) to be easily obtained in the field and easily raised in the laboratory in sufficient number; (iii) to have its genetic composition known; (iv) to have its relative sensitivities to different classes of toxicants known; (v) to be tolerant to the toxicants in question; and (vi) to be representative of its class or phylum. But because it is very difficult to obtain a species that is able to correspond to all the above criteria, in general the chosen species to monitor a certain environment is the one that, fulfilling the most of them, is very abundant or in some way relates to humans by means of the food chain or by means of phylogenetic similarities. Algae, mollusks, fish, small rodents, and humans are some examples of organisms used in environmental monitoring (Solbé, 1993; O'Connor, 1998; Minier *et al.*, 2000; Debenay *et al.*, 2003).

The manifestation of effects caused by a toxicant over a population, community or ecosystem may only occur after several months or years, but on the other hand the cellular or molecular target of that toxicant suffers observable modifications in just few days, hours or even minutes after the exposure takes place. This means that the time interval between the instant of exposure and the instant in which the effects of a toxicant are revealed diminishes from the highest to the lowest biological organization levels (Forbes and Forbes, 1994). To assess biological responses at infra-organism level biological markers, also known as biomarkers, are used (Committee on Biological Markers of the National Research Council, 1987; Schlenk, 1999; Walker *et al.*, 2000).

A biomarker is an induced alteration by one or more toxicants in the components or processes, structures or cellular or biochemical functions, measurable in biological systems or samples (Committee on Biological Markers of the National Research Council, 1987; Klaassen and Watkins III, 2001). Biomarkers may be classified as of exposure, effect or susceptibility. The first type signals the occurrence of exposure to a certain toxicant or class of toxicants, allowing to assess the amplitude of that exposure by means of an estimation of the internal dose either of the toxicant itself or its metabolites in the organism, but they do not provide information on the adverse effects at the organism level. The second type indicates simultaneously the occurrence and the adverse effects of exposure to a certain toxicant and its probable consequences at the population, community or ecosystem level, helping in the qualitative identification of hazards by elucidating possible mechanisms of action. The third type indicates the sensitivity of an organism and allows the evaluation of the inherent or acquired limitation of this organism to the exposure to a certain toxicant (Committee on Biological Markers of the National Research Council, 1987; Depledge, 1994; Connell *et al.*, 1999; Schlenk, 1999; Klaassen and Watkins III, 2001).

From a biological point of view, biomarkers are relevant by contrast to simple analytical measures of concentrations of toxicants in the water, air, soils or food since they allow knowing how the organism suffers under the action of toxicants, and may aid in establishing real thresholds of exposure. They may be used in environmental and health assessment, to identify places and circumstances in which toxicants cause biological effects, in long-term monitoring, to establish base levels for a certain species or area, and to identify environmental deterioration and/or improvement, or in field studies or in the identification of in risk organisms and populations, or in the identification of ways of chemical action of toxicants. Biomarkers may be of molecular, genetic, cellular, tissue or physiological level (Spence *et al.*, 1990; Goksoyr *et al.*, 1996; Hylland *et al.*, 1998; Zaucke *et al.*, 1998; Cajaraville *et al.*, 2000; Karagas *et al.*, 2001; Amaral *et al.*, 2002; Kakkar and Jaffery, 2005).

The pathologically induction of apoptosis by environmental toxicants depending on the chemical species, exposure level and interval, target sites, and energy provision of the cell suggests that apoptosis is not just a physiological mode of cell death (Vaux, 2002). Several publications suggest that when the concentration of a toxicant is high enough to cause irreversible cellular damages or large disorders in the cellular milieu the cell dies by necrosis, without consuming energy, but on the other hand, when concentrations are low it is much more common affected cells to die by apoptosis, which requires consumption of energy. Dioxins, polychlorinated biphenyls (PCBs), cadmium, chromium, and arsenic are just some examples of toxicants capable of inducing apoptosis (Yoo *et al.*, 1997; Kamath *et al.*, 1998; Singh *et al.*, 1999; Feng *et al.*, 2001; Oh and Lim, 2006). Indeed, when inducing apoptosis, low concentration

ns of toxicants are able to initiate cleavage events that may lead to deficient cell cycle sequence, modified mechanisms of homeostasis and repair, scattered and detached structural components, malignant transformation, and targeted death. The detection and quantification of these changes or sequence of changes related to toxicant-induced cell stress and toxicity in several tissues and organs has been suggested to be suitable as a biomarker of exposure to environmental chemical toxicants (Raffray and Cohen, 1997; Mangipudy *et al.*, 1998; Seve *et al.*, 2002).

## Induction of Apoptosis by Organic and Inorganic Toxicants

### Dioxins, PCBs, and PAHs

In numerous cases, the effect an environmental toxicant has on apoptosis is concentration-, cell type-, and cell cycle-dependent. Dioxins, especially 2,3,7,8-Tetrachlorodibenzo-*p*-dioxin (TCDD), PCBs, and PAHs are well-known toxicants that bind to the ligand-activated transcription factor designated aryl hydrocarbon receptor (AHR) (DeVito and Birnbaum, 1995; Yamaguchi *et al.*, 1996, 1997; Lutz *et al.*, 1998; Mann *et al.*, 1999; Nebert *et al.*, 2000; Matzuk, 2001; Maitikanen *et al.*, 2002). TCDD is considered the most toxic of its congeners and one of the most potent environmental toxicant. Adverse effects associated with TCDD exposure include immunotoxicity, reproductive and developmental toxicity, skin lesions, teratogenesis, and carcinogenesis (DeVito and Birnbaum, 1995). Thymic atrophy is perhaps the most documented developmental toxicity associated with TCDD. For instance, prenatal exposure to TCDD has been shown to result in increased apoptosis, mediated by death receptors pathway, in the thymus leading to the atrophy of this organ. This suggests that TCDD may interfere with the T-cell selection processes, ultimately altering the T-cell repertoire and postnatal immune response (Kamath *et al.*, 1998; Camacho *et al.*, 2004). Other studies, using zebrafish embryos, have also found that TCDD causes apoptotic death in the brain (Dong *et al.*, 2001, 2004).

PCBs fit in a large group of persistent chemicals known as halogenated aromatic hydrocarbons, which also comprise the aforementioned dioxins, and have been demonstrated to promote a variety of adverse health effects. PCBs are toxic to several biological systems, including the immune system, reproductive system, nervous system, and endocrine system, have been shown to cause cancer in animals, and are considered potential carcinogens to humans (Safe, 1984, 2000; EPA, 1996; Schantz *et al.*, 2003). Although the commercial production of PCBs was proscribed towards the end of 1970s, they continue to be a health hazard because of their lipophilic nature and their persistence in the environment. These toxicants have been shown to cause DNA laddering and stimulate apoptotic death of neuronal cells, splenocytes, renal tubular cells, thymocytes, and microvascular endothelial cells in a time- and concentration-dependent manner, via the caspases-dependent pathway (Yoo *et al.*, 1997; Lee *et al.*, 2003; Sánchez-Alonso *et al.*, 2003; Lu *et al.*, 2004; Santiago *et al.*, 2006).

PAHs are widely distributed environmental organic toxicants formed mainly by incomplete combustion processes. PAHs are lipophilic and several of them are known or suspected genotoxic carcinogens (WHO, 1998). PAHs such as dimethylben[a]anthracene (DMBA), methylcholanthrene (MCA), and fluoranthene have been implicated in inducing apoptotic cell death in immunodefensive and germ cells. For example, exposure to DMBA has been shown to provoke apoptosis in mice bone marrow pre-B lymphocytes and female germ cells (Lutz *et al.*, 1998; Mann *et al.*, 1999; Matikainen *et al.*, 2002; Yamaguchi *et al.*, 1996, 1997).

## Pesticides, Solvents, and other Organic Compounds

Chronic exposure to toluene and *p*-xylene, which are the most used organic solvents, has been associated with apoptosis that in turn leads to the development and progression of tubulo-interstic ial fibrosis and chronic renal failure (Al-Ghamdi *et al.*, 2004). Another example of induction of apoptosis by organic compounds has also been shown when the cerebellum of weaning male rats underwent increased apoptotic neurodegeneration upon exposure to toluene during gestation (Dalgaard *et al.*, 2001). Combined exposure of rats to DEET (N,N-diethyltoluamide) and permethrin, which are highly used insecticides, has been found not only to induce apoptosis in the brain, as a consequence of the release of cytochrome c from the mitochondrial intermembrane space, but also to induce apoptotic death of testicular germ cells (Abou-Qare and Abou-Donia, 2001; Abou-Donia *et al.*, 2003). The most known and first modern pesticide, DDT [1,1,1-trichloro-2,2-bis(*p*-chlorophenyl)ethane] has been associated to apoptosis in human mononuclear cells and a resulting immunosuppression in exposed children (Pérez-Maldonado *et al.*, 2004). Sulfur dioxide exposure was shown to activate the expression of p53, Bax and caspases-3, and suppress expression of Bcl-2 in the lungs of rats, this way inducing apoptosis (Bai and Meng, 2005). Apoptosis has also been shown to take place in endothelial cells and epidermal keratinocytes, and alveolar cells of the lungs, after exposure to sulfur-mustard [*bis*-(2-chloroethyl)-sulfide] and phosgene (carbonyl chloride), respectively, suggesting it may be used as a biomarker of chemical warfare agent induced toxicology (Dabrowska *et al.*, 1996; Kan *et al.*, 2003; Li *et al.*, 2006). Besides phosgene, another alkane, pristane (2,6,10,14-tetramethylpentadecane) that is used as lubricant and anti-corrosion agent has been found to induce apoptosis both *in vivo* and *in vitro*, and consequently a lupus-like autoimmune syndrome (Calvani *et al.*, 2005).

## Metals: Cadmium

Cadmium (Cd) is a known atmospheric pollutant extensively used in the production of batteries and pigments, and can also be found naturally in the environment. Cd and its compounds are highly toxic even in low concentrations, and bioaccumulate in organisms and ecosystems. The first reported widespread Cd poisoning in the world occurred in Japan, which became known as Itai-Itai disease, when consumption of cadmium-contaminated rice led to kidney failure and softening of the bones. A primary toxic effect of long-term Cd exposure includes kidney injury, involving the formation of a Cd-metallothionein complex (CdMT) in the liver. If transported to the kidney, CdMT undergoes endocytosis and degradation by lysosomes, and $Cd^{2+}$ is liberated, resulting in oxidative stress and apoptosis to the proximal tubules (Hamada *et al.*, 1997; O'brien and Salacinski, 1998). The involvement of Bax and p53 is also suggested to take part in Cd-induced apoptosis (Lag *et al.*, 2002). Cd-induced apoptosis has been shown *in vivo* and *in vitro* in the kidney, liver, brain, testes, and lungs of rodents and/or humans (Xu *et al.*, 1996; Hamada *et al.*, 1997; Habeebu *et al.*, 1998; Lag *et al.*, 2002; Kwon *et al.*, 2003; Wätjen and Beyersmann, 2004; Eichler *et al.*, 2006; Oh and Lim, 2006). For instance, Cd causes oxidative stress-induced neural cell apoptosis (Huang *et al.*, 2006). Besides mammals, Cd-apoptosis has also been found in marine sponges, and trout epithelial cells (Lyons-Alcantara, 1998; Wagner *et al.*, 1998).

## Metals: [Methyl]Mercury

Methylmercury (MeHg), an organometallic cation highly toxic, is an environmental pollutant that causes serious damage to the central nervous system following its consumption in contaminated foods (WHO, 1990). For example, MeHg was the culprit in several massive poisoning episodes that occurred in Minamata and Niigata, both Japanese cities, and Iraq in the 1960s and 1970s. These episodes resulted in paresthesias, loss of physical coordination, difficulty in speech, narrowing of the visual field and blindness, hearing impairment, and even death. Acute and chronic exposure to low concentrations of MeHg has been shown to induce apoptotic death of cells from several biological systems, with special regard to the nervous and immune systems (Kunimoto, 1994; Shenker *et al.*, 1998; Wike *et al.*, 2003). Also, exposure to high doses of MeHg has been found to cause apoptosis but in germ cells of male rats, leading to spermatogenesis impairment (Homma-Takeda *et al.*, 2001). MeHg toxicity is involved in perturbation of mitochondrial function leading to mitochondrial permeability transition, reactive oxygen species generation, and depletion of cellular glutathione and other sulfydryl compounds, leading to apoptosis through the release of cytochrome c (Shenker *et al.*, 1998, 2000; Fonfría *et al.*, 2002).

## Other Metals

Many other metals in elemental form or in compounds have the ability to exert their apoptotic effect, at least partially, by means of generating reactive oxygen species resulting in mitochondrial damage and release of cytochrome c. For instance, exposing human lymphocytes to aluminum or chromium has been shown to induce oxidative stress leading to DNA damage and apoptosis in these cells (Balamurugan *et al.*, 2002; Banasik *et al.*, 2005; Lankoff *et al.*, 2006). Low doses of chromium, and cobalt, have also been found to induce apoptosis in mice macrophages, while high doses cause necrosis (Catelas *et al.*, 2005). Rats administered TBT showing decreased weight of the spleen and thymus because of lymphocyte depletion, by means of TBT-induced oxidative damage and apoptosis (Gennari *et al.*, 2002), and the induction of apoptosis in copper exposed C6/36 cell lines of insect (Raes *et al.*, 2000) are just more evidence of the induction of apoptotic death by metal toxicants.

## CONCLUSION

Regardless of the limited extent of this review concerning to the number of toxicants discussed in opposition to the number of pollutants present in the environment, there is evidence from the abovementioned literature to suggest a pattern in that relevant pollutants may be strong contributors and/or causes to permanent cell damage, tissue or organ dysfunction, and stressed organisms, and that a role for apoptosis in environmental pollutant-induced toxicity is suggested. Therefore, the general association of apoptosis with a large variety of chemical stressors allows characterizing it as a non-specific biological marker of exposure that offers outstanding benefits in assessing the effects of mixtures of toxicants and their metabolites on organism health and in environmental monitoring. These benefits include low variability, high sensitivity, high conservation among phyla, and high adjustability to various environmental compartments (Wagner *et al.*, 1998; Waalkes *et al.*, 2000; Lacorn *et al.*, 2001; Lyons *et al.*, 2004).

Regarding ecotoxicology issues, apoptosis as a biomarker of cell toxicity may be used in diverse animal and non-animal models in order to assess the potential for human exposure and effects. The detection and quantification of apoptotic cell death may also prove useful as part of a set of biomarkers of different categories (e.g. biochemical; physiological; histological) and sensitivities or when a site is under the action of uncharacterized mixtures of toxicants (Depledge, 1994). When considering apoptosis as a biomarker of exposure, it is important to reflect its validity as well as the stability and reversibility of the response (Depledge, 1994; Walker, 1998). For instance, it is likely that apoptosis is valid for the intentions of offering insight into early biochemical episodes and effects of exposure to several types of toxicants, but is also likely that some confounders, operating at the time of assessment, may exist that modulate apoptosis (Lacorn *et al.*, 2001). Biological modulators of apoptosis may include organism age, and seasonal variation in diet and reproduction. Since apoptosis events occur on the order of hours, i.e. are short-lived, and only affect a limited number of cells at a time, the criteria for selection of the right and most useful assay to detect and/or quantify this type of cellular death, and the analysis of the outcomes should consider that apoptosis involves a continuum of early and late occurrences that are not coordinated in time or have the same duration. In addition, it is vital to recognize what molecular controls and events connect cell stress to the irreversible process of apoptosis. As mentioned before, apoptosis occurs in normal physiological conditions as a way of controlling the size of cell populations, therefore when apoptosis is reduced or stimulated or when cellular proliferation is excessive the organism may be under pathological conditions, induced perhaps by chemical stressors.

In present days, there are many environmental monitoring programs and point studies that use apoptosis as one of several biomarkers, included in a suite of other biological markers, and that suggest it as a very useful one for biomonitoring purposes (Kamath *et al.*, 1998; Wagner *et al.*, 1998; Frenzilli *et al.*, 2004; Amaral and Rodrigues, 2005). The detection and quantification of apoptosis especially in association with others biomarkers, such as heat shock proteins or metallothioneins, may be considered an important biomarker for *in vivo* exposure to toxicants. Indeed, the ease, fastness, and low cost of apoptosis measurements, including flow cytometry, TUNEL assay, and Comet assay when compared to organism level ecological indices makes apoptosis an important mainstream biomarker that can be used in biomonitoring. In addition, there are several assays to measure apoptosis in which blood cells are used and do not require the organism to be dead. Thus, detecting and characterizing toxicant-induced apoptosis may be considered a general biomarker of response because this type of cell death is expressed in multiple cell systems, is conserved phylogenetically, is influenced by environmentally significant levels of toxicants, and signals an early response to chemical environmental exposure indicative of the status of stress of the organism.

## REFERENCES


Abou-Donia, MB; Suliman, HB; Khan, WA; Abdel-Rahman, AA. Testicular germ-cell apoptosis in stressed rats following combined exposure to pyridostigmine bromide, N,N-diethyl-m-toluamide (DEET), and permethrin. *Journal of Toxicology and Environmental Health A*, 2003, 66(1), 57-73.

Abou-Qare, AW; Abou-Donia, MB. Combined exposure to DEET (N,N-diethyl-m-toluamide) and permethrin-induced release of rat brain mitochondrial cytchrome c. *Journal of Toxicology and Environmental Health A*, 2001, 63(4), 243-252.

Al-Ghamdi, SS; Raftery, MJ; Yaqoob, MM. Organic solvent-induced proximal tubular cell apoptosis via caspases-9 activation. *Environmental Toxicology and Pharmacology*, 2004, 16, 147-152.

Amaral, AFS, Rodrigues, AS. Metal accumulation and apoptosis in the alimentary canal of *Lumbricus terrestris* as a metal biomarker. *BioMetals*, 2005, 18(3), 199-206.

Amaral, AF; Alvarado, N; Marigómez, I; Cunha, R; Hylland, K; Soto, M. Autometallography and metallothionein immunohistochemistry in hepatocytes of turbot (*Scophthalmus maximus* L.) after exposure to cadmium and depuration treatment. *Biomarkers*, 2002, 7(6), 491-500.

Bai, J; Meng, Z. Effects of sulfur dioxide on apoptosis-related gene expressions in lungs from rats. *Regulatory Toxicology and Pharmacology*, 2005, 43, 272-279.

Balamurugan, K; Rajaram, R; Ramasami, T; Narayanan, S. Chromium(III)-induced apoptosis of lymphocytes: Death decision by ROS and Src-family tyrosine kinases. *Free Radical Biology and Medicine*, 2002, 33(12), 1622-1640.

Banasik, A; Lankoff, A; Piskulak, A; Adamowska, K; Lisowska, H; Wojcik, A. Aluminum-induced micronuclei and apoptosis in human peripheral-blood lymphocytes treated during different phases of the cell cycle. *Environmental Toxicology*, 2005, 20(4), 402-406.

Beeby, A. What do sentinels stand for? *Environmental Pollution*, 2001, 112, 285-298.

Cajaraville, MP; Bebianno, MJ; Blasco, J; Porte, C; Sarasquete, C; Viarengo, A. The use of biomarkers to assess the impact of pollution in coastal environment of the Iberian Peninsula: a practical approach. *Science of the Total Environment*, 2000, 247, 295-311.

Calow, P. General principles and overview. In: Calow P, editor. *Handbook of Ecotoxicology*. Oxford: Blackwell Scientific Publications; 1994; 1-5.

Calvani, N; Caricchio, R; Tucci, M; Sobel, ES; Silvestris, F; Tartaglia, P; Richards, HB. Induction of apoptosis by the hydrocarbon oil pristane: Implications for pristane-induced lupus. *Journal of Immunology*, 2005, 175(7), 4777-4782.

Camacho, IA; Nagarkatti, M; Nagarkatti, PS. Evidence for induction of apoptosis in T cells from murine fetal thymus following perinatal exposure to 2,3,7,8-tetrachlorodibenzo-*p*-dioxin (TCDD). *Toxicological Sciences*, 2004, 78, 96-106.

Catelas, I; Petit, A; Vali, H; Fragiskatos, C; Meilleur, R; Zukor, DJ; Antoniou, J; Huk, OL. Quantitative analysis of macrophage apoptosis vs. necrosis induced by cobalt and chromium ions in vitro. *Biomaterials*, 2005, 26(15), 2441-2453.

Committee on Biological Markers of the National Research Council. Biological markers in environmental health research. *Environmental Health Perspectives*, 1987, 74, 3-9.

Connell, D; Lam, P; Richardson, B; Wu, R. Introduction to Ecotoxicology. Oxford: Blackwell Science Ltd.; 1999.

Cunningham, WP; Cunningham, MA; Saigo, B. Environmental Science. A Global Concern. 8th ed. New York: McGraw-Hill; 2005.

Dabrowska, MI; Becks, LL; Lelli Jr., JL; Levee, MG; Hinshaw, DB. Sulfur mustard induces apoptosis and necrosis in endothelial cells. *Toxicology and Applied Pharmacology*, 1996, 141, 568-583.

Dalgaard, M; Hossaini, A; Hougaard, KS; Hass, U; Ladefoged, O. Developmental toxicity of toluene in male rats: effects on semen quality, testis morphology, and apoptotic neurodegeneration. *Archives of Toxicology*, 2001, 75, 103-109.

Debenay, JP; Carbonel, P; Morzadec-Kerfourn, MT; Cazaubon, A; Denefle, M; Lezine, AM. Multi-bioindicator study of a small estuary in Vendee (France). *Estuarine Coastal and Shelf Science*, 2003, 58(4), 843-860.

Depledge, MH. The rational basis for the use of biomarkers as ecotoxicological tools. *In*: Fossi, MC; Leonzio, C, editors. *Nondestructive Biomarkers in Vertebrates*. Boca Raton: Lewis Publishers; 1994; 271-295.

Depledge, MH; Weeks, JM; Bjerregaard, P. Heavy metals. In: Calow P, editor. *Handbook of Ecotoxicology*. Oxford: Blackwell Scientific Publications; 1994; 79-105.

DeVito, MJ; Birnbaum, LS. Dioxins: model chemicals for assessing receptor mediated toxicity. *Toxicology*, 1995, 102, 115-123.

Dong, W; Teraoka, H; Kondo, S; Hiraga, T. 2, 3, 7, 8-tetrachlorodibenzo-p-dioxin induces apoptosis in the dorsal midbrain of zebrafish embryos by activation of arylhydrocarbon receptor. *Neuroscience Letters*, 2001, 303(3), 169-172.

Dong, W; Teraoka, H; Tsujimoto, Y; Stegeman, JJ; Hiraga, T. Role of aryl hydrocarbon receptor in mesencephalic circulation failure and apoptosis in zebrafish embryos exposed to 2,3,7,8-tetrachlorobenzo-*p*-dioxin. *Toxicological Sciences*, 2004, 77, 109-116.

Eichler, T; Ma, Q; Kelly, C; Mishra, J; Parikh, S; Ransom, RF; Devarajan, P; Smoyer, WE. Single and combination toxic metal exposures induce apoptosis in cultured murine podocytes exclusively via the extrinsic caspase 8 pathway. *Toxicological Sciences*, 2006, 90(2), 392-399.

EPA. PCBs: Cancer Dose-Response Assessment and Application to Environmental Mixtures. Washington: EPA; 1996.

EPA. Biological indicators of watershed health. 2005. Available from: http://www.epa.gov/bioindicators/.

Feng, ZM; Xia, YJ; Tian, DF; Wu, KK; Schmitt, M; Kwok, RK; Mumford, JL. DNA damage in buccal epithelial cells from individuals chronically exposed to arsenic via drinking water in Inner Mongolia, China. *Anticancer Research,* 2001, 21(1A), 51-57.

Fonfría, E; Dare, E; Benelli, M; Suñol, C; Ceccatelli, S. Translocation of apoptosis-inducing factor in cerebellar granule cells exposed to neurotoxic agents inducing oxidative stress. *European Journal of Neuroscience*, 2002, 16(10), 2013-2016.

Forbes, VE; Forbes, TL. Ecotoxicology in Theory and Practice. Ecotoxicology Series II. London: Chapman & Hall; 1994.

Frenzilli, G; Scarcelli, V; Barga, I; Nigro, M; Förlin, L; Bolognesi, C; Sturve J. DNA damage in eelpout (*Zoarces viviparus*) from Göteborg harbour. *Mutation Research*, 2004, 552, 187-195.

Gabrielides, GP. Pollution of the Mediterranean Sea. *Water Science Technology*, 1995, 32(9-10), 1-10.

Gennari, A; Bol, M; Seinen, W; Penninks, A; Pieters, R. Organotin-induced apoptosis occurs in small $CD4^{+}CD8^{+}$ thymocytes and is accompanied by an increase in RNA synthesis. *Toxicology*, 2002, 175, 191–200.

Goksoyr, A; Beyer, J; Egaas, E; Grosvik, BE; Hylland, K; Sandvik, M; Skaare, JU. Biomarker responses in flounder (*Platichthys flesus*) and their use in pollution monitoring. *Marine Pollution Bulletin*, 1996, 33(1-6), 36-45.

Habeebu, SSM; Liu, J; Klaassen, CD. Cadmium-induced apoptosis in mouse liver. *Toxicology and Applied Pharmacology*, 1998, 149, 203-209.

Hamada, T; Tanimoto, A; Sasaguri, Y. Apoptosis induced by cadmium. *Apoptosis*, 1997, 2, 359–367.

Homma-Takeda, S; Kugenuma, Y; Iwamuro, T; Kumagai, Y; Shimojo, N. Impairment of spermatogenesis in rats by methylmercury: involvement of stage- and cell- specific germ cell apoptosis. *Toxcicology*, 2001, 169, 25-35.

Huang YH, Shih CM, Huang CJ, Lin CM, Chou CM, Tsai ML, Liu TP, Chiu JF, Chen CT. Effects of cadmium on structure and enzymatic activity of Cu,Zn-SOD and oxidative status in neural cells. *Journal of Cellular Biochemistry*, 2006, 98(3), 577-589.

Hwang, JS; Kobayashi, C; Agata, K; Ikeo, K; Gojobori, T. Detection of apoptosis during planarian regeneration by the expression of apoptosis-related genes and TUNEL assay. *Gene*, 2004, 333, 15-25.

Hylland, K; Nissen-Lie, T; Christensen, PG; Sandvik, M. Natural modulation of hepatic metallothionein and cytochrome P4501A in flounder, *Platichthys flesus* L. *Marine Environmental Research*, 1998, 46(1-5), 51-55.

Jouany, JM. Environment and human health. *Ecotoxicology and Environmental Safety*, 1998, 40, 2-3.

Kakkar, P; Jaffery, FN. Biological markers for metal toxicity. *Environmental Toxicology and Pharmacology*, 2005, 19, 335-349.

Kamath, AB; Nagarkatti, PS; Nagarkatti, M. Characterization of phenotypic alterations induced by 2,3,7,8-tetrachlorodibenzo-*p*-dioxin on thymocytes *in vivo* and its effect on apoptosis. *Toxicology and Applied Pharmacology*, 1998, 150, 117-124.

Kan, RK; Pleva, CM; Hamilton, TA; Anderson, DR; Petrali, JP. Sulfur mustard-induced apoptosis in hairless guinea pig skin. *Toxicologic Pathology*, 2003, 31(2), 185-190.

Karagas, MR; Stukel, TA; Morris, JS; Tosteson, TD; Weiss, JE; Spencer, SK; Greenberg, ER. Skin cancer risk in relation to toenail arsenic concentrations in a US population-based case-control study. *American Journal of Epidemiology*, 2001, 153(8), 559-565.

Kerr, JFR. History of the events leading to the formulation of the apoptosis concept. *Toxicology*, 2002, 181-182, 471-474.

Klaassen, CD; Watkins III, JB. Manual de Toxicología. La Ciencia Básica de los Tóxicos. 5th ed. México: McGraw-Hill Interamericana; 2001.

Kunimoto, M. Methylmercury induces apoptosis of rat cerebellar neurons in primary culture. *Biochemical and Biophysical Research Communications*, 1994, 204, 310-317.

Kwon, KY; Jang, JH; Kwon, SY; Cho, CH; Oh, HK; Kim, SP. Cadmium induced acute lung injury and TUNEL expression of apoptosis in respiratory cells. *Journal of Korean Medical Science*, 2003, 18, 655-662.

Lacorn, M; Piechotta, G; Simat, TJ; Kammann, U; Wosniok, W; Lang, T; Muller, WEG; Schroder, HC; Jenke, HS; Steinhart, H. Annual cycles of apoptosis, DNA strand breaks, heat shock proteins, and metallothionen isoforms in dab (*Limanda limanda*): influences of natural factors and consequences for biological effect monitoring. *Biomarkers*, 2001, 6(2), 108-126.

Lag, M; Westly, S; Lerstad, T; Bjornsrud, C; Refsnes, M; Schwarze, PE. Cadmium-induced apoptosis of primary epithelial lung cells: involvement of Bax and p53, but not of oxidative stress. *Cell Biology and Toxicology*, 2002, 18, 29-42.

Landis, WG; Yu, MH. Introduction to Environmental Toxicology: Impacts of Chemicals Upon Ecological Systems. Boca Raton: Lewis Publishers; 1995.

Lankoff, A; Banasik, AB; Duma, A; Ochniak, E; Lisowska, H; Kuszewski, T; Gozdz, S; Wojcik, A. A comet assay study reveals that aluminium induces DNA damage and inhibits the repair of radiation-induced lesions in human peripheral blood lymphocytes. *Toxicology Letters*, 2006, 161(1), 27-36.

Lee, YW; Park, HJ; Son, KW; Hennig, B; Robertson, LW; Toborek, M. 2,2',4,6,6'-pentachlorobiphenyl (PCB 104) induces apoptosis of human microvascular endothelial cells through the caspases-dependent activation of CREB. *Toxicology and Applied Pharmacology*, 2003, 189, 1-10.

Li, WL; Hai, CX; Liang, X; Zhang, XD; Chen, HL; Qin, XJ; Liu, R; He, W; Wang, P; Li, B. Apoptosis of ATII cells in mice induced by phosgene. *Inhalation Toxicology*, 2006, 18, 71-77.

Lu, Z; Lee, EY; Robertson, LW; Glauert, HP; Spear, BT. Effect of 2,2',4,4',5,5'-hexachlorobiphenyl (PCB-153) on hepatocyte proliferation and apoptosis in mice deficient in the p50 subunit of the transcription factor NF-kB. *Toxicological Sciences*, 2004, 81, 35-42.

Lutz, CT; Browne, G; Petzold, CR. Methylcholanthrene causes increased thymocyte apoptosis. *Toxicology*, 1998, 128, 151–167.

Lyons, BP, Bignell, J; Stentiford, GD; Feist, SW. The viviparous blenny (*Zoarces viviparus*) as a bioindicator of contaminant exposure: application of biomarkers of apoptosis and DNA damage. *Marine Environmental Research*, 2004, 58, 757-761.

Lyons-Alcantara, M; Mooney, R; Lyng, F; Cottell, D; Mothersill, C. The effects of cadmium exposure on the cytology and function of primary cultures from rainbow trout. *Cell Biochemistry and Function*, 1998, 16, 1-13.

Maitikanen, TM; Moriyama, T; Morita, Y; Perez, GI; Korsmeyer, SJ; Sherr, DH; Tilly, JL. Ligand activation of the aromatic hydrocarbon receptor transcription factor drives Bax-dependent apoptosis in developing fetal ovarian germ cells. *Endocrinology*, 2002, 143(2), 615-620.

Mangipudy, RS; Rao, PS; Andrew, A; Bucci, TJ; Witzmann, FA; Mehendale, HM. Dose-dependent modulation of cell death: apoptosis versus necrosis in thioacetamide hepatotoxicity. *International Journal of Toxicology*, 1998, 17, 193-211.

Mann, KK; Matulka, RA; Hahn, ME; Trombino AF; Lawrence, BP; Kerkvliet, NI; Sher, DH. The role or polycyclic aromatic hydrocarbon metabolism in dimethylbenz[*a*]anthracene-induced pre-B lymphocyte apoptosis. *Toxicology and Applied Pharmacology*, 1999, 161, 10-22.

Matzuk, MM. Eggs in the balance. *Nature Genetics*, 2001, 28, 300-301.

Minier, C; Levy, F; Rabel, D; Bocquené, G; Godefroy, D; Burgeot, T; Leboulenger, F. Flounder health status in the Seine Bay. A multibiomarker study. *Marine Environment Research*, 2000, 50, 373-377.

Moriarty, F. Ecotoxicology. The Study of Pollutants in Ecosystems. London: Academic Press; 1983.

Nebert, DW; Roe, AL; Dieter, MZ; Solis, WA; Yang, Y; Dalton, TP. Role of the aromatic hydrocarbon receptor an [*Ah*] gene battery in the oxidative stress response, cell cycle control, and apoptosis. *Biochemical Pharmacology*, 2000, 59, 65-85.

Newman, MC; Lawrence, CA; Unger, MA. *Fundamentals of Ecotoxicology*. 2$^{nd}$ ed. Boca Raton: Lewis Publishers; 2002.

O'Brien, P; Salacinski, H. Evidence that the reactions of cadmium in the presence of metallothionein can produce hydroxyl radicals. *Archives of Toxicology*, 1998, 72, 690-700.

O'Connor, TP. The NOAA National Status and Trends program. *Marine Pollution Bulletin*, 1998, 37(1-2), 3-5.

Oh, S-H; Lim S-C. A rapid and transient ROS generation by cadmium triggers apoptosis via caspase-dependent pathway in HepG2 cells and this is inhibited through *N*-

acetylcysteine-mediated catalase upregulation. *Toxicology and Applied Pharmacology*, 2006, 212, 212-223.

Pérez-Maldonado, IN; Díaz-Barriga, F; Fuente, H; González-Amaro, R; Calderón, J; Yáñez, L. DDT induces apoptosis in human mononuclear cells in vitro and is associated with increased apoptosis in exposed children. *Environmental Research*, 2004, 94, 38-46.

Phillips, DJH. Quantitative Aquatic Biological Indicators: Their Use to Monitor Trace Metal and Organochlorine Pollution. London: Applied Science Publishers Ltd.; 1980.

Raes, H; Braekman, BP; Criel, GRJ; Rzeznik, U; Vanfleteren, JR. Copper induces apoptosis in Aedes C6/36 cells. *Journal of Experimental Zoology*, 2000, 286(1), 1-12.

Raffray, M; Cohen, GM. Apoptosis and necrosis in toxicology: a continuum or distinct modes of cell death? *Pharmacology & Therapeutics*, 1997, 75(3), 153-177.

Safe, S. Polychlorinated biphenyls (PCBs) and polybrominated biphenyls (PBBs): biochemistry, toxicology, and mechanism of action. *Critical Reviews in Toxicology*, 1984, 13(4), 319-395.

Safe, SH. Endocrine disruptors and human health - Is there a problem? An update. *Environmental Health Perspectives*, 2000, 108(6), 487-493.

Sánchez-Alonso, JA; López-Aparicio, P; Recio, MN; Pérez-Albarsanz, MA. Apoptosis-mediated neurotoxic potential of a planar (PCB 77) and a nonplanar (PCB 153) polychlorinated biphenyl congeners in neuronal cell cultures. *Toxicology Letters*, 2003, 144, 337-349.

Santiago, MF; Pérez-Reyes, PL; López-Aparicio, P; Recio, MN; Pérez-Albarsanz, MA. Differential effects of PCBs on the induction of apoptosis machinery and PKCα translocation in rat renal tubular cell cultures. *Toxicology Letters*, 2006, 163, 91-100.

Schantz, SL; Widholm, JJ; Rice, DC. Effects of PCB exposure on neuropsychological function in children. *Environmental Health Perspectives*, 2003, 111(3), 357-376.

Shenker, BJ; Guo, TL; Shapiro, IM. Low-level methylmercury exposure causes human T-cells to undergo apoptosis: evidence of mitochondrial dysfunction. *Environmental Research*, 1998, 77, 149–159.

Shenker, BJ; Guo, TL; Shapiro, IM. Mercury-induced apoptosis in human lymphoid cells: Evidence that the apoptotic pathway is mercurial species dependent *Environmental Research*, 2000, 84(2), 89-99.

Schlenk, D. Necessity of defining biomarkers for use in ecological risk assessments. *Marine Pollution Bulletin*, 1999, 39(1-12), 48-53.

Seve, M; Chimienti, F; Favier A. Rôle du zinc intracellulaire dans la mort cellulaire programmée. *Pathologie Biologie*, 2002, 50, 212-221.

Singh, J; Pritchard, DE; Carlisle, DL; Mclean, JA; Montaser, A; Orenstein, JM; Patierno, SR. Internalization of carcinogenic lead chromate particles by cultured normal human lung epithelial cells: formation of intracellular lead-inclusion bodies and induction of apoptosis. *Toxicology and Applied Pharmacology*, 1999, 161, 240-248.

Solbé, JFLG. Freshwater fish. In: Calow P, editor. *Handbook of Ecotoxicology*. Oxford: Blackwell Scientific Publications; 1993; 66-82.

Spence, SK; Hawkins, SJ; Santos, RS. The mollusc *Thais haemastoma* - an exhibitor of 'imposex' and potential biological indicator of tributyltin pollution. *Marine Ecology*, 1990, 11(2), 147-156.

Truhaut, R. Ecotoxicology: objectives, principles and perspectives. *Ecotoxicology and Environmental Safety*, 1977, 1, 151-173.

Vaux, DL. Apoptosis and toxicology—what relevance? *Toxicology*, 2002, 181-182, 3-7.

Waalkes, MP; Fox, DA; States, JC; Patierno, SR; McCabe, MJ. Metals and disorders of cell accumulation: Modulation of apoptosis and cell proliferation. *Toxicological Sciences*, 2000, 56(2), 255-261.

Wagner, C; Stefen, R; Koziol, C; Batel, R; Lacorn, M; Steinhart, H; Simat, T; Müller, WEG. Apoptosis in marine sponges: a biomarker for environmental stress (cadmium and bacteria). *Marine Biology*, 1998, 131, 411-421.

Walker, CH. Biomarker strategies to evaluate the environmental effects of chemicals. *Environmental Health Perspectives*, 1998, 106(S2), 613– 620.

Walker, CH; Hopkin, SP; Sibly, RM; Peakall, DB. Principles of Ecotoxicology. 2nd ed. London: Taylor and Francis; 2000.

Wätjen, W; Beyersmann, D. Cadmium-induced apoptosis in C6 glioma cells: influence of oxidative stress. *BioMetals*, 2004, 17, 65-78.

WHO. Environmental Health Criteria 101: Methylmercury. Geneva: WHO; 1990.

WHO. Non-heterocyclic Polycyclic Aromatic Hydrocarbons. Geneva: WHO; 1998.

Wilke, RA; Kolbert, CP; Rahimi, RA; Windebank, AJ. Methylmercury induces apoptosis in cultured rat dorsal root ganglion neurons. *NeuroToxicology*, 2003, 24, 369-378.

Wright, DA; Welbourn, P. Environmental Toxicology. New York: Cambridge University Press; 2000.

Xu, C; Johnson, JE; Singh, PK; Jones, MM; Yan, H; Carter, CE. In vivo studies of cadmium-induced apoptosis in testicular tissue of the rat and its modulation by a chelating agent. *Toxicology*, 1996, 107, 1-8.

Yamaguchi, K; Near, R; Shneider, A; Cui, H; Ju, ST; Sherr, DH. Fluoranthene-induced apoptosis in murine T cell hybridomas is independent of the aromatic hydrocarbon receptor. *Toxicology and Applied Pharmacology*, 1996, 139, 144-152.

Yamaguchi, K; Matulka, RA; Shneider, AM; Toselli, P; Trombino, AF; Yang, S; Hafer, LJ; Mann, KK; Tao, XJ; Tilly, JL; Near, RI; Sherr, DH. Induction of preB cell apoptosis by 7,12-dimethylbenz[*a*]anthracene in long-term primary murine bone marrow cultures. *Toxicology and Applied Pharmacology*, 1997, 147, 190-203.

Yoo, BS; Jung, KH; Hana, SB; Kim, HM. Apoptosis-mediated immunotoxicity of polychlorinated biphenyls (PCBs) in murine splenocytes. *Toxicology Letters*, 1997, 91, 83–89.

Zaucke, F; Zöltzer, H; Krug, HF. Dose-dependent induction of apoptosis or necrosis in human cells by organotin compounds. *Fresnius' Journal of Analytical Chemistry*, 1998, 361, 386-392.

Zhang, J; Xu, M. Apoptotic DNA fragmentation and tissue homeostasis. *TRENDS in Cell Biology*, 2002, 12(2), 84-89.

# INDEX

## B



## C

## D

## E

## H



## I

## J



## K



## L



## M

## N

## O



## P

## R



## S

## T

## U



## V



## W



## X



## Y



## Z